THE PRODUCT MANAGEMENT PLAYBOOK

CREATE, SHIP, AND OPTIMIZE WINNING PRODUCTS

Julia Barham

Rosenfeld

NEW YORK 2026

"Pragmatic, comprehensive, and refreshingly useful—the book every aspiring product manager needs, and every experienced one will wish they'd had sooner."

—Vijay Raghavendra
Chief Technology and Product Officer at GEICO

"Product management is evolving fast and becoming more critical than ever in an AI-driven world. This book is an absolute must-read for any aspiring or ambitious product manager."

—Derek Schwartz
Head of Data Science and Analytics at Intuit—Identity Platform

"This book doesn't just teach product management theory—it hands you the exact tools, templates, and plays to tackle every phase of the life cycle with confidence."

—Melinda Seibold
SVP at TD Insurance

"This book is the structured, confidence-building guide I wish had existed earlier in my product management career."

—Mike Belsito
Director of Mind the Product at Pendo.io

"An excellent playbook for product managers looking to sharpen their toolkit and level up."

—Rich Mironov
Author of *Money Stories* and *The Art of Product Management*

"This is the practical, no-nonsense handbook for product managers at every level. Read it to transform complexity into clear, actionable steps."

—Clay Johnson
VP of Product at TD, formerly Meta and Capital One

"This book is less theory, more hard-earned truth from someone who's actually been in the trenches... essential reading for anyone making the leap into product management."

—Puneet Chopra
Senior Director of Engineering, formerly Amazon and Green Dot

"This approachable guide to product management shares valuable advice on a wide breadth of topics that PMs need to be successful."

—Dan Olsen
Product management consultant, trainer, and author of
The Lean Product Playbook

"A clear, practical guide that helps product managers move from ideas to evidence and execution with confidence."

—David J. Bland
Founder at Precoil and co-author of *Testing Business Ideas*

"Finding product-market fit is difficult. This book gives you the concepts, actions, and advice to make it possible."

—Chetan Kandhari
Chief Innovation and Digital Officer at Nationwide

"Finally, the practical strategies of product management in one place. If you had one book to propel your product career forward, this is it."

—Jamie Anderson
Brand and Customer Experience Design Executive

"The skills and perspectives that make a great product manager are hard to catalog, let alone acquire. Julia Barham has written a book I can finally recommend to any aspiring or ambitious PM to show them the ropes—a complete tour through the full breadth of product craft."

—Christian Crumlish
Author of *Product Management for UX People* and
Director of Product Management for Kind Systems

"A smart tour of essential topics across contemporary product management, written by a guiding mentor who's here to help you assess gaps and apply proven ways of moving forward."

—Jake Burghardt
Author of *Stop Wasting Research,*
former Principal Product Manager at Amazon

"Product management is more important now than ever before. This book is a foundational reference that will make any product person better."

—Josh Seiden
Author of *Outcomes Over Output*

The Product Management Playbook
Create, Ship, and Optimize Winning Products
By Julia Barham

Rosenfeld Media, LLC
125 Maiden Lane
New York, New York 10038
USA

On the Web: www.rosenfeldmedia.com
Please send errata to: errata@rosenfeldmedia.com

Publisher: Louis Rosenfeld
Managing Editor: Marta Justak
Interior Layout: Danielle Foster
Cover Design: Heads of State
Illustrator: Julia Barham
Indexer: Marilyn Augst
Proofreader: Sue Boshers

ISBN: 1- 959029-07-X
ISBN 13: 978-1-959029-07-6
LCCN: 2026930500

For my grandparents, Julio and Julia Rodríguez,
whose love and wisdom carry on

For my family, whose support means everything

And for Eric and Sally—my heart and home

HOW TO USE THIS BOOK

Who Should Read This Book?

This book is for anyone who turns messy problems into real products and then lives with the results. Whether you're an aspiring product manager, an experienced veteran, stuck in a "feature factory," or a people leader of product teams, this book is the one-stop reference guide to help you and your team get "unstuck." That said, read this book if you are:

- A **product manager** (or similar role) ready to round out your core competencies, build impactful products, and prepare for leadership roles.
- A **product leader** seeking a structured guide to coach your team and break product work into core concepts, methods, and reusable templates.
- A **digital professional** or **career-changer** curious about modern product management practices and what the PM role really entails.

If you're a current product manager, think of this book as your product coach. Take the skills assessment to pinpoint gaps and opportunities in your personal growth journey. Then lean on the concepts and playbook methods to tackle different stages of the product development process with confidence. Throughout the book, you'll uncover techniques to build high-impact products that are not only feasible but also desirable *and* viable. Plus, you'll learn exactly what C-level leaders and partners want to see from you, including specific strategies for leading cross-functional teams and influencing stakeholders. And for those seeking career longevity, you'll learn tips to build career resilience, prepare for product leadership roles, and shine during interviews!

If you're a product leader, think of this book as the assistant coach on your team. Use the skills assessment to understand gaps and strengths of direct reports. And instead of throwing new product managers into the deep end with minimal training, share the foundational concepts, techniques, and reusable templates with your team so they can start off on the right foot.

If you're an aspiring product manager or digital professional, read this book to demystify the wide-ranging role of product managers in the wild. You'll discover the everyday duties of PMs, skills every product builder should master, and proven product management methods you can use (even if your company hasn't formally adopted the product operating model).

What's in This Book?

Like the role of a product manager, this book covers a lot of ground. It's designed to level up your competencies across strategic thinking, product discovery, delivery, and leadership. While the advice is tailored for corporate environments, the methods are universal and will benefit those in nonprofits and other industries as well.

Here's what you'll find inside:

- **A roadmap to guide your growth:** Chapter 2, "The Effective Product Manager's Toolkit, offers a skills assessment to baseline your competencies today, while Chapters 11, "Mastering Leadership and Influence," and 12, "The Future of Project Management and You," offer tips for interviewing, advice for those on the path to leadership, and strategies for career longevity. Throughout the book, I've also curated additional resources and thought leaders you should follow for continued learning on special topics.
- **21 step-by-step product management methods:** Whether you're building an MVP from scratch, scaling an in-market solution, inheriting a fledgling product, or anywhere in between, there's a method that can help. There's even a method to clarify the business goals so you can spend more time on work that matters and less time debating priorities.
- **Reusable templates and examples:** To save you from reinventing the wheel, you'll encounter dozens of discovery and delivery guides you can use in everyday work.
- **8 interviews featuring expert advice:** Learn what separates elite PMs from the rest, straight from renowned thought leaders, C-level executives, and cross-functional experts who work with world-class product teams.

What Comes with This Book?

This book's companion website (https://rosenfeldmedia.com/books/product-management-playbook/) contains the book's diagrams and other illustrations—all available under a Creative Commons license (when possible) for you to download and include in your own presentations.

FREQUENTLY ASKED QUESTIONS

What makes this book different from other product books?

Most product books today do one of two things: they stay at the level of theory and principles, or they plant a flag on a single topic and go deep. Both have value, but few acknowledge the reality for everyday practitioners: managing a portfolio of features in different stages of maturity, trying to balance discovery and delivery simultaneously, working without durable design and data partners, and dealing with needy stakeholders.

This book is meant to give you something you can use on Monday morning when a new problem hits your inbox. Wherever you are in the product life cycle, Chapters 3–10 have a playbook method that can help focus your team and help you march forward with confidence.

I'm not a product manager—is this book still for me?

Absolutely. "Product manager" is a title, but product thinking often shows up in dozens of roles across companies because it spans five critical domains, which are covered in Chapter 1, "The Truth About Product Management." Designers, engineers, strategists, product owners, business analysts, and founders are all doing product work to some degree. Even teams that don't follow the product operating model still need to define problems, create solutions, and achieve product-market fit.

This book uses the term *PM* as shorthand, but it's written for anyone building technology-enabled solutions to meet customer needs and business goals. If that's the work you do, this book is for you. Plus, if you plan to interview for product manager roles, Chapter 12 "The Future of Product Management and You" includes interview strategies and resources.

I'm an experienced product manager. Do I need to read this?

Yes, for three reasons. First, the field of product management is a moving target. Experience doesn't equal longevity. This book covers enduring principles and modern technologies, tools, and ways of approaching product craft today. Everyone has to stay current.

Second, *thousands* of experienced product managers have blind spots. Many were stuck in delivery roles for years and then expected to become strategists overnight. Chapters 3–7 will help them level up discovery skills, strategic thinking, and stakeholder influence techniques. Other folks spent years cooking up "strategy," but don't have enough execution experience to know how the two connect. Chapters 6–10 will help these readers go deep on the importance of using data and technology as a strategic lever, how to lead delivery teams, and how to assess product-market fit.

Third, if you want to move into product leadership, you'll be expected to coach less-experienced PMs. To do this, you can use the skills assessment in Chapter 2, "The Effective Product Manager's Toolkit," to evaluate your team and develop individualized coaching plans. You can also use the principles in Chapter 11, "Mastering Leadership and Influence," to help teammates build a strong brand and prepare for promotion.

Does this book cover AI tools and AI product management concepts?

Yes! Throughout the book, I encourage teams to supercharge the product development process using AI productivity tools, with one critical caveat: AI tools should supplement, not substitute core product thinking. Chapter 6, "How: Defining the Approach," takes a deeper dive, covering the key technology concepts every PM should understand about AI today, including five specific technologies and six skills every product manager should develop when working on AI-based products. Notably, this book also addresses two emerging areas that every product manager and business leader should understand: how to evaluate third-party AI solutions and how to practice responsible AI standards.

Is there a single methodology that works best in product management?

No. Product management varies greatly by company, team size, product maturity, culture, and org design. A PM working for a large enterprise will have a different experience than a PM operating at a five-person startup. That said, Chapter 2 covers four competencies that are required in almost every context (strategic thinking, discovery skills, delivery skills, and leadership behaviors) and common methodologies used by product teams today.

CONTENTS

FOREWORD

A question I keep coming back to, as someone who writes a lot about product management and talks to a highly diverse sample of people in the industry, is whether all the advice out there is actually helping.

There's a lot of talk of "product thinking," yet when you try to figure out what that actually means you're either sold a workshop or end up navigating a whole lot of hand-waving and "you'll know it when you see it." Or books that paint an oversimplified view of the craft, without caveats, without appreciation for how hard the work actually is. Or "vibes and thinking" advice from "top-tier" companies where the story on the inside is always messier than the sound bites on the podcasts.

I ask myself this, too. Am I helping or hurting? Product management is deeply contextual. Not everyone works at Google, Stripe, Anthropic, or Linear. And not every nondigital-product-selling company is a bureaucratic mess where you can't get your reps and do good work. How do you actually hone your craft when it's so hard for any one person to explore every context?

What people actually need is a mentor. But not just anyone. Someone who has worked across genuinely different environments: not just Silicon Valley startups or just big enterprises, not just B2C or B2B, not just "empowered" product orgs or highly regulated, change-averse ones. Someone who gives you specific, actionable plays but doesn't sugarcoat how long it takes to get good at this, tells you when something is 101 and when it will take two years. Someone who built a personal playbook by actually using it, in practice, across those different worlds.

Julia is that person. She has led products at a subscription media company, built experimentation culture at a major bank, launched a digital-first insurance model from scratch, and scaled acquisition portfolios serving millions of customers. That range matters. It's the reason the advice in this book holds up, even when your context doesn't match the Silicon Valley default.

Product management is in a strange place right now. With AI, we may be on the edge of a Cambrian explosion of better products and tighter connections to value. At the same time (in 2026), the drumbeat of layoffs and hot takes about everything being "dead" or "changing forever" can make the field feel precarious. The advice I keep coming back to, whether someone is new to product management or experienced, is to focus on getting your reps in, expand your thinking beyond the narrow set of high-profile tech companies, become highly adaptive, and keep learning by doing. That is easier said than done when you feel overwhelmed, when the landscape keeps shifting, and when you're not sure which voice to trust.

What Julia has done here is something I didn't think was possible without writing a thousand pages or staying only an inch deep. She balances "do this now" with "but keep in mind." She gives you concrete plays and then tells you, plainly, that the perfect framework doesn't exist and that context will change the answer. From the opening self-assessment, the tone is set: I've been there, I've been evolving this playbook, and I have just enough ideas and perspective to keep you moving forward. Above all, it is obvious that Julia cares. About practical advice, about people, and about not sugarcoating any of it.

That's what makes this an anti-gatekeeping guide. Not a map that pretends the territory is simple, but a mentor who has been through enough of it to help you navigate.

—John Cutler
Author of *The Beautiful Mess* newsletter

INTRODUCTION

Like many people before me, I fell into product management by accident. I was guessing about ways to boost retention for a digital subscription product and overheard colleagues mumbling about *jobs-to-be-done* and *design thinking*. After digging in, the lightbulb went off! Instead of starting with new features in mind, why not begin with customer research and map out exactly what users need? Then I could define solutions that customers valued and meet business goals.

That aha moment ignited an ultramarathon of learning for me: hundreds of podcasts, dozens of books, framework deep-dives, and career-defining mentorships all helped me become fluent across the product life cycle. Over a decade, I built a personal treasure trove of templates and resources, so I'd never have to reinvent the wheel again when starting a new assignment. I shared them with peers and used them to train product teams from scratch, but I realized these tools could do more: they could bridge the massive confidence gap that many product managers (PMs) face today.

In fact, the more PMs I spoke with, the more I noticed a concerning pattern: despite leading high-stakes work, most of them had *not* received product management training. Instead, they were thrown into the deep end with minimal coaching and forced to piecemeal lessons from podcasts, articles, books, and Reddit threads—just like I was doing over a decade ago!

That's when another lightbulb went off. Why not write the playbook I wish I'd had early in my career? One that admits honest realities of life in the trenches and offers practical ways for PMs to create structure out of ambiguity.

This playbook does just that. It's for the ambitious product management professional who wants to create high-impact solutions but needs an extra boost of confidence and guidance. It's a distillation of selected problem-solving frameworks and methods you can use to build products that keep customers happy and businesses healthy.

Many of the techniques stand on the shoulders of thought leaders who came before me, and while this book packs a punch, it's not the gospel. In fact, there's no "right" way to build products. Product management is a craft that never stops evolving.

So, consider this book your menu of options. When you're stuck, facing a new challenge, or need inspiration, try out a method or template listed here to find a way forward. Use whatever works for your situation and adapt or change what doesn't. Harness that special superpower every product manager needs—the ability to *figure it out*.

I'm confident you've got this. Now let's get to work.

CHAPTER 1

The Truth About Product Management

Ask a product manager to describe their job, and you'll hear classic analogies, like *orchestrators, translators*, and *team quarterbacks*. Ask their design and engineering partners? You'll get cheeky responses like *cat-herders* and *chaos-wranglers*. Or there's what your family calls you... a *project manager*. (I promise, it's not the same thing, Mom!)

But honestly speaking, product management is a practice and titles don't matter. Outcomes matter. And the product manager's job is to accomplish one outcome: define and ship useful products that create value for customers and companies.

So, product managers do whatever it takes. They wear multiple hats, learn many skills, and get really nosey—developing expertise across business, the market, design, data, and technology domains to define the North Star vision and product strategy. Along the way, they build bridges by bringing partners and experts together to create alignment and momentum. And they provide strategic leadership and structured planning to guide teams across the product development process—from problem exploration to solution design, delivery, and continuous optimization.

Are you considering a career in product management? It's a lot harder than it looks. Fair warning: your day-to-day duties and blockers will fluctuate depending on your solution and your company's maturity. Constraints are a daily reality in this field, so remove "easy" from your vocabulary entirely. On good days, people will still call you a project manager, and on tough days, you might call yourself crazy. But if you love rallying teams, making sense of ambiguity, and building useful products that create value for customers and companies, product management can be the most energizing role in the world.

NOTE PRODUCT MANAGERS VS. PROJECT MANAGERS

Product managers are responsible for defining why a solution matters, who benefits from it, and what the solution requires before shifting into when and how it's delivered. *Project managers* usually focus on when and how work is delivered after the solution is defined. That said, lines can blur at some companies.

The Product Manager's Role Explained

This book is for product managers (otherwise known as PMs) who define, build, and ship *technology-enabled* solutions, including digital products, services, and experiences.[1] As the connective tissue behind solutions that define our everyday lives and drive business growth, PMs are dynamic team players that tackle a staggering list of responsibilities. Figure 1.1 shows the duties many PMs shoulder today.

The Product Manager's Duties

- Translate market research, customer insights, and business goals into world-class products that satisfy customers and deliver measurable value.
- Establish a product vision, product strategy, and milestone-based roadmap that aligns stakeholders and defines success through key metrics.
- Decompose roadmaps into prioritized feature sets that consider business impact, feasibility, and customer benefit.
- Author clear product requirements and maintain a product backlog (including epics, features, and user stories) to guide teams building technology-enabled solutions.
- Lead cross-functional partners across the product development process, including problem definition, research, prototyping, prioritization, design, development, testing, launch, and in-market optimization.
- Influence and maintain alignment across business, design, engineering, data, marketing, compliance, and operations partners during product development.
- Manage complex trade-offs involving scope, resources, and timelines to ensure high-quality execution.
- Scale and optimize in-market solutions using evidence-based decision-making while balancing innovation, incremental enhancements, and production support.
- Pivot or sunset solutions that no longer meet strategic goals or customer needs.
- Monitor customer, platform, and business data and synthesize insights into actionable steps that feed strategic decisions and product improvements.
- Communicate product health and performance updates to stakeholders and ensure transparency and alignment on progress toward business goals.

FIGURE 1.1
Duties and methods will vary across companies. Chaos-wrangling is universal.

1 While products, services, and experiences technically vary in meaning, this book covers universal aspects of product management and uses the terms *product* or *solution* interchangeably to keep things simple.

In an ideal situation, the product manager's role closely resembles this description, but what's on paper often doesn't reflect reality. For example, not all companies clearly articulate their business goals, which makes translating those goals into solutions very difficult. Also, many PMs lack durable access to user research, design, data, and marketing experts. Even worse, some businesses are still learning why product discovery and customer research matter. When reality and expectations don't align, every product manager harnesses an essential superpower: tackling ambiguity.

CLARITY IS THE JOB: THE FOUR TYPES OF AMBIGUITY PRODUCT MANAGERS TACKLE EVERYDAY

Product management is uniquely an end-to-end process. Few roles span the entire product development process from initial insight to post-launch optimization. End-to-end ownership is rewarding, but it also creates pressure to have all the answers. Because PMs sit at the intersection of many stakeholders and handoffs, your teams will look to you for direction when chaos starts to rise. To lead effectively, you must learn to tackle these four types of everyday ambiguity with a steady hand:

- **Problem ambiguity:** Business stakeholders speak in "features" while users speak in "symptoms." In a sea of conflicting feedback, PMs don't just listen—they synthesize. They use data and sharp judgment to define what really matters, why it matters, and who to optimize for.
- **Priority ambiguity:** On Monday, you have one clear goal and by Tuesday three "fires" landed on your desk. When everything seems urgent, the PM operates as a filter for team focus, deciding what's important to tackle first and why.
- **Solution ambiguity:** A brilliant strategy with poor execution leads nowhere. Great PMs partner closely with design and engineering teams to understand the art of the possible. And they craft holistic requirements to ensure that the team builds a solution that meets customer *and* business needs.
- **Organizational ambiguity:** Fuzzy reporting structures, multiple handoffs, and shifting dependencies are a rite of passage for PMs in large organizations. The best PMs don't wait for blockers to emerge. Instead, they build informal networks, call their contacts, and influence decision-makers to clear organizational blockers before they stall progress.

The Core Domains of Product Management

Shipping world-class products requires multidisciplinary skills and a collaborative mindset. To be successful, product management demands expertise across five core domains: the market, your business, design, technology, and data (see Figure 1.2). By fusing and synthesizing insights across each of these areas, PMs develop a product sense: the ability to craft technology-enabled solutions for products that satisfy customers and build durable businesses. Knowledge across these domains also makes you, the product manager, a stronger collaborator and partner when leading cross-functional teams. Throughout this book, you will find concepts, methods, and resources to level up your domain knowledge across each area.

FIGURE 1.2

Expertise in one or two of these domains makes a transition into product management easier.

Domain expertise doesn't just help you make better decisions—it helps you master a vital skill in everyday product management: influence-based leadership. Yes, new PMs should build knowledge in each domain to develop a strategic point-of-view, but success often hinges on how well you use that knowledge to align disparate roles and voices into a singular mission. Depending on context, domain expertise can help you be the shield for your team (protecting them from distractions or chaos) or the sword that cuts through indecision by communicating in a way that resonates with your stakeholders.

Flavors of Product Management

There's no ideal path into product management. The field is a mosaic of professionals that arrive from diverse backgrounds. Many PMs start in product management with a T-shaped skill set, meaning they arrive with deep expertise in one or two domains and broad competence in the others. Depending on the company's needs and individual expertise, they either step into generalist or specialist roles. Generalist PMs are like a Swiss-army knife—they have a broad range of knowledge and can work across the entire product life cycle. Specialist PMs often excel in specific technologies or business outcomes. Table 1.1 shows several common flavors of product management roles today. (Keep in mind, these are not mutually exclusive; some PM roles are a hybrid version of these.)

TABLE 1.1 COMMON TYPES OF PRODUCT MANAGERS

Type of PM	Areas of Focus	Examples
PMs That Build Core Products and Experiences	B2C products B2B/enterprise products B2B2C products	Banking mobile app CRM tool Payment solutions
PMs That Specialize in Specific Domains	Innovation Growth/optimization Data products AI/ML products	Startup products Activation experiments A/B testing platform Personalization engines
PMs That Enable Others to Build and Create Solutions	Platform PMs Technical PMs Internal tools PMs	Integration platforms External/internal APIs Native MarTech products

Like the role itself, product team structures vary widely. Depending on the company, you might find yourself aligned to business outcomes, customer segments, user journeys, specific features, or technical architecture boundaries.[2] As an individual PM, you won't be dictating team topology or org charts; however, understanding how to navigate your organization is a critical skill. Having awareness of the company's structure and reporting relationships will help you quickly identify and collaborate with partners. Start building your network now!

2 Although PMs and teams will vary, this book covers the universal aspects of product management and common responsibilities on a product team.

THE ORIGIN STORY OF PRODUCT MANAGEMENT

If you're wondering where modern product management comes from, here's an abridged history lesson that starts in the 20th century.

1920s

To combat product commoditization, Procter & Gamble began market research to understand how and why customers used their household products. In 1931, they invented brand management teams to ensure that consumer insights shaped product, packaging, and distribution strategies.

1950s

The *Toyota Production System*, developed in the mid-1900s, transformed the manufacturing process through two core principles: *jidoka* (automation with a human touch) and *just-in-time* (continuous production flows that minimize waste). This product management strategy scaled high-quality, low-cost production by blending machine automation with human wisdom.

1990s

Toyota's approach inspired Lean, a business management framework focused on value stream efficiency and the elimination of unnecessary activities. Businesses, nonprofits, and government functions embraced this new way of working to develop, deliver, and manage goods and services.

2000s

Influenced by modern business management and manufacturing methods, seventeen experts published the "Agile Manifesto" in 2001 to overcome rigid software development practices. The manifesto principles inspired several product management methodologies rooted in user-centered development, cross-functional teams, and continuous delivery. The shift in product development practices transformed early product managers from isolated requirements owners to a central team figure focused on aligning why, who, what, how, and when across the team.

2020s

Today, product management is a high-leverage, interdisciplinary function at thousands of digitally-transformed companies. PMs are no longer an extension of technology teams; in many cases, they now report to product teams and chief product officers. In the modern era, PMs operate at the intersection of customers, business, design, and technology to continuously define, build, and ship products that are useful, feasible, and viable.

The Biggest Myths in Product Management

Some of you reading this book are current PMs trying to level up your game. Others are considering the leap into product management. Regardless of your stage, there are two big misconceptions that skew expectations and stall growth for product managers worldwide. Let's debunk these myths right now so you can move forward with a clearer sense of purpose and empowerment.

Myth 1: The Myth You're Told

Contrary to what you read and hear, product managers are *not* mini-CEOs. Yes, there are *some* parallels, given the ownership mindset and leadership skills required to be a PM, but overall, the PM's job comes with less razzle-dazzle and a lot less authority than a CEO. It's true that in some small companies, many product managers often execute founder-driven ideas. However, while the PM may have a broader responsibility for the customer experience, they usually have less influence in the product's vision and strategic direction than the co-founder. In large companies, many PMs lead small, cross-functional teams with a well-defined scope, but they are removed from broader vision and strategy conversations. Across all companies, it's rare for product managers to own the P&L (profit and loss) responsibility. For example, CEOs can choose their teams, dictate resources, and control the budget. Product managers, however, can't do any of that.

Instead of a mini-CEO, here's a less glamorous, but more accurate metaphor: product managers are the *glue* of organizations. They are masters at unifying, filling gaps, and combining discrete insights together to establish product sense. They align customer needs to business goals and operate at the intersection of five large domains—business, customer, technology, data, and design—to create valuable solutions. Along the way, they bring together dozens of partners, subspecialties, and dependent teams, ensuring that everyone understands the core vision and works toward the same goals. And they do it without any formal authority.

Myth 2: The Myth You Tell Yourself

Despite what job descriptions say, many early career PMs do not participate explicitly in vision setting and strategy formation. This is especially common when PMs own specific features or capabilities in a large product. In this environment, it's easy to believe that leaders create the vision and strategy, and you just deliver it. (And yes, many times product leaders *are* responsible for a cohesive product vision across a large portfolio of products and services.)

But the myth is believing that your scope is *not* worthy enough for a vision and a strategy, too. Even if your boss doesn't explicitly say so, no product scope is too small to think strategically. Take authentication, for example. If you own the log-in experience of an app, what's stopping you from imagining the North Star vision and crafting a strategy to make that a world-class solution? Nothing. The same thinking applies to other examples, too. Whenever you own a specific problem area, you should find inspiration, create a vision and strategy, and proactively share it. You don't need to wait for someone to ask you—start practicing now, especially if you see a product leader role in your future.

Why Product Managers Matter More Than Ever

Recent advances, mainly driven by AI technologies, are starting to blur lines across product thinkers, product builders, and everyone in between. Today, it's easier than ever to create a product and launch a solution quickly. But speed often sacrifices quality, and developing a solution instantly doesn't mean it's inherently useful and valuable. Remember, product management is a practice that requires strategic thinking across multiple domains, not just execution chops.

Companies will always need individuals who think deeply about market opportunities, customer problems, and user needs. They need people who will envision brand-new value propositions and build solutions in previously unimaginable ways. And of course, they'll need thinkers and doers who can navigate organizational politics and keep teams moving toward a shared vision. These skills will become more critical as automation eliminates mundane tasks.

Yes, AI is reshaping and accelerating parts of the product development life cycle in powerful ways, but even AI product teams require human-in-the-loop checkpoints. To be sure, the future still needs product people. Truth-seekers and truth-tellers. Strategists and executors. The glue that holds everything together.

Tackling the Confidence Gap PMs Feel Today

Right now, thousands of product managers are fumbling their way through product work every day, and honestly, who can blame them? Job descriptions don't match reality, teams are spread thin, and many PMs work for companies that are attempting half-baked transformations. As more companies have gone *all-in on digital,* many people flipped into product management job titles with zero coaching. Zilch! And when PMs change companies, most are thrown into the deep end on day 1 without any training. To make matters more concerning, many PMs are panicking about career longevity, given the rise of AI.

The confidence gap many product managers face today is real. But you can still become an exceptional product manager despite these constraints. You just need a realistic mindset about your responsibilities, a toolkit of best practices, and a ton of grit. That said, becoming an exceptional product manager is a marathon, *not* a sprint. Set your expectations accordingly. Even with great coaching, it's impossible to become a solid product manager in less than two years.

To develop real product sense, you need experience with different solutions, teams, and product life cycles. (And product life cycles can move slower than you'd expect.)[3] Sure, you can attend bootcamps or conferences, but hands-on exposure is where real learning happens. It helps you recognize patterns across people, products, and processes that lead to success or failure.

But with any race, success is *so much easier* when you've got support and advice. Across each chapter, you'll acquire skills and methods to tackle every aspect of the product life cycle with examples, expert advice, templates, and practical strategies. You'll learn tips to develop a stronger brand with partners, adopt a player-coach management style, and prepare for a journey toward product leadership. By the end, you'll have an end-to-end view of the product manager's responsibilities and techniques for tackling whatever assignment comes next... and tackling it with confidence.

So, consider this book your product management marathon coach. Your training starts now.

---

3 Product life cycles' timeframes vary across companies and often depend on a mix of factors including company culture, funding, resource constraints, regulatory challenges, and of course, market insights.

CHAPTER 2

The Effective Product Manager's Toolkit

I spent my early years in product management building digital subscription products for a small market. I loved the industry and work, but I wanted something more—for example, experience shipping products for millions of users. After a year of hunting, I landed a gig that *really* pushed me out of my comfort zone. The new job sent me straight into delivery mode with a sprint team building enterprise capabilities. Without deep technical acumen, I fumbled my way through architecture conversations, product requirements, and rapid release cycles. After four months of subpar performance, my confidence hit a low point.

With coaching and support from peers, I decided to press on week after week. I began asking senior PMs questions like, "How *exactly* do you run your team meetings?," "How would you decompose this problem?," and "What *is* vertical slicing?" I also reworked my schedule to protect time for deep thinking and found an architecture mentor. With support, I adapted. As my skills developed, my confidence returned. Fast forward two years, and I was leading multiple sprint teams, delivering impact consistently, and even training new product managers.

Of course, the story didn't end there. Over time, I led innovation work with executives, developed internal platforms for colleagues, ran a sales funnel for a Fortune 100 company, and built a product team from scratch. But every career evolution followed a similar pattern as the first: new role, big feelings, perseverance, and then a major growth spurt! I'd be lying if I said each change was easy (it wasn't), but I learned to trust the process. Across many growth spurts, I gained experience across the entire product life cycle and began my journey as a product leader.

Why share this story? Because well-rounded product managers don't emerge by playing it safe. They seek out the unknown—new skills, fresh ways of working, and different org designs. And they apply new insights to build better products, lead with courage, and shape their future. Whether you take big career swings or small ones, your longevity in product management boils down to three important questions:

- **Are you motivated to learn?** Your learning journey requires self-awareness, endless rounds of iteration, and motivation to gain new skills. But the payoff is worth it.
- **Are you willing to be uncomfortable?** Excellence requires growth, and growth requires discomfort. Sometimes the best career opportunities are the ones that scare you.

- **Are you ready to be patient?** Each milestone took *years* of learning, exposure, and yes, some tears. When the doubt creeps in, trust the process.

If you're ready to say yes to all three questions, let's start your growth journey together right now.

Assessing Your Product Skills

You don't need a stint in Silicon Valley to become a world-class product manager; you just need to be *deliberate* about honing your craft. Many PMs sharpen their skills by changing companies or industries while others grow in place by finding mentors and following thought leaders. Whichever path you choose, progress depends on your openness and intention.

So, in the spirit of self-awareness, it's time to create a baseline of where you stand today. Grab a pen and get started filling out Table 2.1. Review each skill and check the box that best describes your current ability level. (If you have minimal or no experience with a skill and require coaching, select *rookie*. If you have meaningful experience with a skill and can demonstrate independent proficiency, select *practitioner*. If you're highly experienced with a skill and could train others how to do it, select *expert*. When finished, count the total number of check marks in each column and list each column's total in the final row.)

TABLE 2.1 THE PRODUCT MANAGEMENT SKILLS ASSESSMENT

Skill Description	I'm a rookie and need help doing this.	I'm a practitioner and can do this without help.	I'm a n expert and can teach others how to do this.
SECTION 1			
Identifies product life cycle stages and adapts goals and activities at each stage.			
Builds domain expertise in customer needs, business models, market dynamics, and emerging trends.			
Elicits and uses company strategy to shape team activities and product direction.			

continues

Skill Description	I'm a rookie and need help doing this.	I'm a practitioner and can do this without help.	I'm an expert and can teach others how to do this.
Defines problem/ opportunity statements to scope and guide product development activities.			
Assesses competitors and identifies strategies to differentiate solutions.			
Evaluates product opportunities across desirability, feasibility, and viability data and uses insights to inform next steps.			
Establishes a product vision and strategy that defines the North Star and strategic bets for new and existing solutions.			
Leverages hypothesis-driven frameworks to manage risk and uncertainty.			
Recommends and manages strategic pivots or sunsets a product when needed.			
SECTION 2			
Develops research goals for generative, evaluative, and summative research.			
Translates customer research into user need statements that align team efforts.			
Synthesizes market, business, design, data, and technology insights into product direction.			
Prioritizes problems and solutions using desirability, viability, and feasibility inputs.			

continues

Skill Description	I'm a rookie and need help doing this.	I'm a practitioner and can do this without help.	I'm an expert and can teach others how to do this.
Prototypes solutions, tests concepts, and runs experiments to de-risk assumptions before committing significant resources.			
Elicits solution architecture, technology options, and feasibility assessments to inform product direction.			
Elicits data and analytics strategy and uses it to inform product direction.			
Defines holistic product requirements (business, user, functional, nonfunctional).			
SECTION 3			
Forms product roadmap, including objectives, scope, and success metrics over a given timeframe.			
Decomposes roadmap into iterations and maintains healthy backlog to optimize delivery velocity.			
Decomposes requirements into user stories with sprint team.			
Supports delivery activities during software development (dependency management, acceptance testing, launch readiness).			
Leads or participates in go-to-market planning (pricing, marketing strategy, operational readiness).			
Implements techniques to de-risk product launches.			

continues

Skill Description	I'm a rookie and need help doing this.	I'm a practitioner and can do this without help.	I'm an expert and can teach others how to do this.
Establishes feedback loops to monitor product performance and adapt roadmap.			
Delivers quality improvements alongside enhancements (eliminates defects, avoids tech debt, manages production support).			
Assesses and achieves product-market fit for new and existing solutions.			
Proactively prepares for scale by implementing best practices and standards.			
Continuously assesses funnel, customer behavior, and user research insights to inform next steps.			
SECTION 4			
Presents product plans to leadership and stakeholders and secures buy-in.			
Leverages and adapts appropriate methodologies to guide discovery and delivery activities (i.e., Jobs-to-Be-Done, design thinking, Agile, etc.).			
Leverages strong partnerships with design teams across product development life cycle (PDLC).			
Leverages strong partnerships with engineering teams across PDLC.			

continues

Skill Description	I'm a rookie and need help doing this.	I'm a practitioner and can do this without help.	I'm an expert and can teach others how to do this.
Leverages strong partnerships with data science and data engineering teams across PDLC.			
Negotiates trade-offs and resolves conflicts among stakeholders.			
Models high-performing team behaviors, holds team accountable, and practices team recognition.			
Audits and improves the team's way of working to improve efficiency.			
Crafts and communicates product performance updates to stakeholders.			
Leads a portfolio of multiple products or solutions at once.			
Solicits and uses partner feedback to improve personal performance.			
Models continuous learning, leadership maturity, and resilience.			
Totals			

Before we talk results, let's unpack that assessment. The survey you just completed covers 40 real-life skills you must master to become a well-rounded product manager. Each skill maps to one of four competency areas that define product management craft—strategic thinking, product discovery, delivery, and leadership. Here's how each section of the skills assessment maps to the competencies:

- **Section 1—Strategic Thinking:** How well the PM understands their customer, company, and industry and translates domain mastery into strategic direction.

- **Section 2—Discovery Skills:** How well the PM uncovers, analyzes, and synthesizes complex business or customer problems, identifies potential solutions, and validates the desirability, feasibility, and viability of those solutions.
- **Section 3—Delivery Skills:** How well the PM drives execution and go-to-market activities across teams and manages solutions in-market, including optimization, de-risking techniques, and product pivots.
- **Section 4—Leadership Behaviors:** How well the PM aligns and partners with stakeholders at every level, including functional experts, peers, and senior leadership. This area requires exceptionally strong collaboration, leadership, and communication skills.

With these competency areas in mind, let's talk about results. If you scored 32 or higher in the third column (expert), you're probably an exceptional product manager already with years of diverse career experiences and strong leadership potential. However, some of you may notice entire sections where you marked rookie because of limited career experiences, even if you're a clear practitioner or expert in other competencies. Ultimately, if your score doesn't reflect expert-level skills across the board, fear not! It's totally normal—it just means you still have room to grow.

How to Use This Assessment

Self-awareness is power. Use your assessment results to bridge skills gaps and guide development conversations. Whether you're considering side quests in your current role or eyeing a bigger career move, this book will guide you across each competency area and the entire product development process. Chapters 3–7 focus on high-value discovery skills you'll use during problem exploration and early solution design. Chapters 8–10 teach practical ways to lead delivery, go-to-market planning, and manage in-market solutions. Chapters 11–12 offer strategies to build leadership maturity, grow your influence, and build career resilience. Throughout, you'll learn concepts and concrete knowledge to sharpen your strategic thinking competency, too. So, the next time you land in unknown territory, you'll have context, advice, and specific methods to get *unstuck*.

New to Product Management? Start Here

Every new and aspiring PM needs a solid foundation in how real-world product managers *think*. The rest of this chapter will introduce *six foundational concepts* every PM can use to anchor themselves in everyday work. This includes three conceptual mental models (or thinking frameworks) that help PMs understand and diagnose a product's health and potential before building a strategy and roadmap. These three mental models are:

- The product life cycle
- Product-market fit and why you need it
- The trifecta of assessing products—desirability, viability, and feasibility (DVF framework)

The remaining concepts cover processes and techniques you can use to build alignment with teams that design, ship, and optimize products. Simply put, they help you *activate* product thinking. They include:

- Product design methodologies you can use to frame and solve problems
- Software development methodologies teams use to ship solutions
- Product artifacts PMs create to set direction for the team

Together, these six concepts provide the foundational bedrock to start tackling uncertainty and guide product work, regardless of industry or company size. Study them first to start your transition from a task manager to a real product thinker.

The Product Life Cycle

The product life cycle, shown in Figure 2.1, represents different stages of a solution's lifespan, starting with the original business "hunch," moving through the product's introduction to the marketplace, shifting to growth and scale, and ending with the product's decline or deprecation. As a PM, you'll need to manage products in various stages of maturity, so it's critical to understand which phase you're leading and how to craft strategies based on that stage. Here are the five phases of the product life cycle and their objectives at each stage.

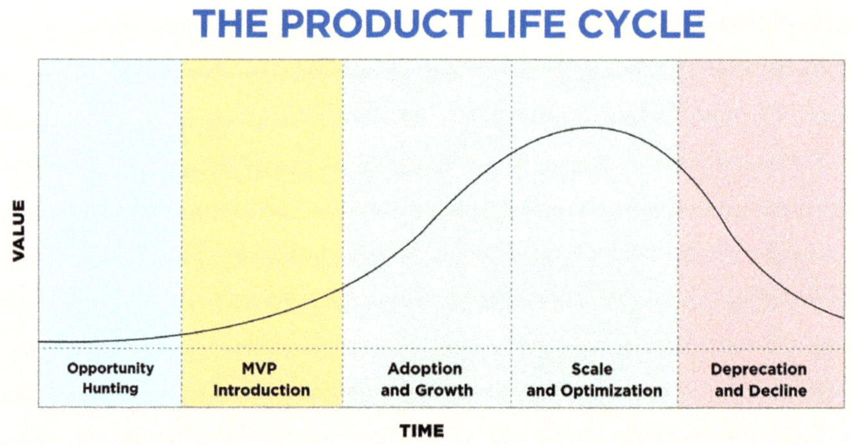

FIGURE 2.1
Determine which phase fits best with your current product assignment.

Opportunity Hunting

- **How to identify this phase:** Your company is "hunting" for new opportunities, but the problem and solution hypothesis aren't clear. This is common among early innovation teams, strategy groups, or when emerging tech (like AI) arrives. It often starts from two directions—either your company observes a pain point and commissions research that indicates a market gap, or your company starts with a solution in search of a problem.
- **What PMs must do:** Transform extreme ambiguity into a clear strategic bet. First, align stakeholders on the company's goals, success criteria, strategic beliefs, and the ideal target market. Then deeply explore the problem space. Synthesize customer research, market analysis, and emerging conditions into a clear opportunity or problem statement and begin generating and testing solution hypotheses.

Minimum Viable Product (MVP) Introduction

- **How to identify this phase:** Your company has a well-researched product concept, but you need in-market feedback to validate that the solution delivers value to the market and company. In this phase, the company attempts to build and launch a minimum version of the product that can deliver enough value to assess market demand and viability.

- **What PMs must do:** Launch an MVP to assess product-market fit and identify a path toward sustainable business value. Use in-market data to establish insights about target segments, value proposition elements, benefits, pricing, and distribution strategies *before* investing in growth. Continue enhancing the product for growth or pivot core hypotheses to achieve product-market fit.

Adoption and Growth

- **How to identify this phase:** Your product or solution clearly meets the needs of a specific market segment, you have paying customers already, organic growth is rising, and your company wants to reach more customers while maintaining and improving product performance.
- **What PMs must do:** Implement strategies that retain existing users and acquire more customers. Use customer feedback to evolve the product, improve onboarding, and increase stickiness for current users without sacrificing quality. Invest in product-led or sales-led growth tactics, experience optimization, and find new distribution channels to acquire new customers.

Scale and Optimization

- **How to identify this phase:** Your customer base is no longer growing rapidly, but you still have sustainable market share, and the product is delivering revenue or expected value based on company goals.
- **What PMs must do:** Emphasize operational excellence and maximize the product's value by reducing costs and increasing productivity. Optimize key customer journeys and improve efficiency while avoiding customer churn.

Deprecation Decline

- **How to identify this phase:** Your solution no longer aligns to the company's long-term goals, and the cost to maintain outweighs financial value. The product may be losing market share because stronger alternatives exist.
- **What PMs must do:** Form a go-forward strategy to sunset the product while protecting current customers and the company from "harm." Help customers convert seamlessly to a new alternative.

Did you notice how PMs must shift their focus and strategy to deliver unique outcomes across each phase? That's why this job is so tough. You can spend a lifetime developing skills and strategies to be successful at each phase. That said, some PMs find themselves drawn to specific parts of the product life cycle and build careers specializing in certain phases. If that's you, fantastic! Specialists are absolutely needed. But if you're a junior PM and you see a future in product leadership roles, seek experiences that offer exposure to different parts of the life cycle. Doing so will prepare you for one of the toughest parts of leadership—managing a portfolio with *multiple* products in *different* phases.

Product-Market Fit Explained

Product-market fit. It's catchy, it's exciting, it's what product managers rave about...but what exactly is it?

Product-market fit (PMF) is an outcome that occurs when a company delivers a solution to a market segment and two things happen: First, the solution satisfies the market segment's needs so effectively that customers don't just adopt it—they stay.

Second, the solution delivers measurable value back to the company, usually in the form of customer growth or revenue. Without PMF, you're fighting uphill economics. With high acquisition costs, low retention, and weak monetization, you'll burn through resources faster than you can create value.

The absolute best framework for understanding PMF components comes from Dan Olsen's *The Lean Product Playbook*[1] where he establishes the Product-Market Fit Pyramid (shown in Figure 2.2) to help innovators understand how to bridge the problem-space and solution-space. I've added another element at the bottom of this pyramid to remind PMs that company objectives provide the foundation for all strategic product decisions.

Achieving PMF requires coordination and accountability across many teams, including business, product, tech, design, sales, operations, and beyond. However, product and business leaders often hold PMs accountable because they are the "glue" across the team. I'll cover tips to assess and achieve PMF in Chapter 10, "Managing Products In-Market."

1 Dan Olsen, *The Lean Product Playbook: How to Innovate with Minimum Viable Products and Rapid Customer Feedback* (Wiley, 2015).

THE PRODUCT-MARKET FIT PYRAMID

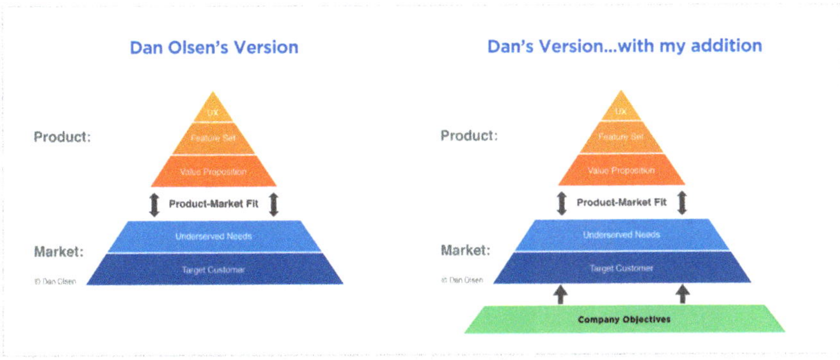

IMAGE COURTESY OF DAN OLSEN

FIGURE 2.2

Use the Product-Market Fit Pyramid to establish alignment between problem definition and solution design.

For now, keep this in mind: no scope is too small to consider product-market fit. I've used this pyramid countless times to build or reverse engineer strategic assumptions, craft strategies, and communicate the building blocks of PMF to stakeholders. Whether you're leading a fully-featured product with millions of customers, or you're working on a small feature, use Dan's pyramid to pressure test your thinking and strengthen PMF.

PRO TIP USE THE PRODUCT-MARKET FIT PYRAMID TO EVALUATE PRODUCT PERFORMANCE

When performance data doesn't match company expectations, bring out the Product-Market Fit Pyramid and document the strategic choices made for each layer. Next, fold in product data and customer feedback. Does the data indicate a core assumption is wrong? If so, begin research and iteration for that layer.

Alternatively, when you're just starting a new initiative, you can use the Product-Market Fit Pyramid to consider what your next steps are for the research and development process. Just think of the pyramid levels like building blocks and use those to anchor your team's goals and next steps.

The DVF Framework

IDEO, a well-known design firm, created the "DVF" (desirability, viability, feasibility) framework in the early 2000s as a helpful model to mitigate risks during product development. Many product teams use the DVF mental model to reduce the chance of shipping a solution that doesn't sell, doesn't return value back to the organization, or isn't sustainable to maintain. The criteria include:

- **Desirability:** The degree to which users want and need your product.
- **Viability:** The degree to which your solution makes good business sense and justifies the cost of producing it.
- **Feasibility:** The degree to which your company can build, distribute, and maintain the solution given financial, technical, and operational resources, along with legal or regulatory constraints.

PMs can use the DVF mental model in everyday work. For example, each time you evaluate a new solution hypothesis, receive a feature request, or prioritize backlog items, ask yourself these questions before committing development resources.

- **Desirability:** Do users want or need this product, feature, or solution? Will they actually use it?
- **Viability:** Does the proposed solution make good business sense? What specific financial benefit does it bring to our company? For example, will it generate revenue, reduce costs, or deliver other measurable benefits that justify the cost of producing it and maintaining it?
- **Feasibility:** Do we have the financial, technical, and operational capabilities needed to deliver and maintain this solution? What are the limitations or dependencies we need to overcome?

In some organizations, the DVF framework is formalized into stage-gates or product reviews where teams present data-driven research and compelling evidence before unlocking funding or launching a new product. Figure 2.3 includes the types of analysis business leaders might expect during these checkpoints, although methods vary by organization. These checkpoints help companies manage risk and allocate resources to opportunities with the highest chance of success. PMs often work with their direct leaders and partners across design, engineering, finance, data, marketing, and operations to complete these assessments.

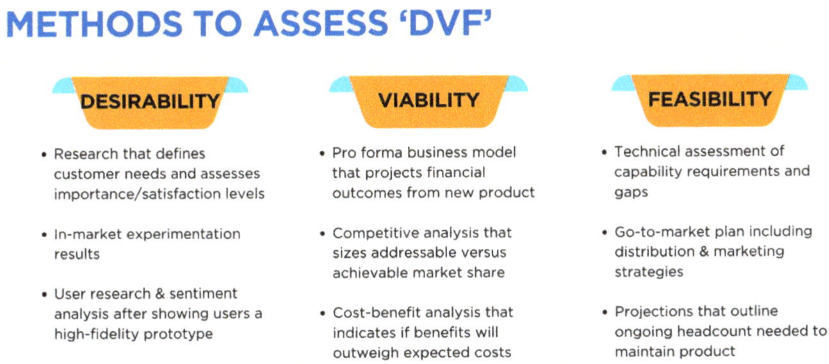

FIGURE 2.3
Some companies use these methods to make go/no-go decisions before the product launches.

Product Design Methodologies

The product life cycle, Product-Market Fit Pyramid, and DVF framework are mental models that help you decide *what* to focus on when crafting your product strategy, but they don't offer a step-by-step process that explains *how* to build or execute on that strategy. That's where product design methodologies help.

Product design methodologies give companies a shared language and structured approach when identifying problems or opportunities, designing solutions, and iterating on solutions in-market. Most companies anchor their product development life cycle (PDLC) steps around one or more of these established product design methodologies:

- **Design thinking:** Popularized by IDEO and Stanford in the early 2000s, design thinking is a human-centered approach many businesses use to innovate (see Figure 2.4). Design thinking is a useful framework to uncover the customers' point-of-view, build customer empathy on teams, and assess desirability of new solutions. Because the process is heavily focused on customer-centric innovation, it's not always intuitive for organizations that differentiate products based on technology breakthroughs or business model pivot.

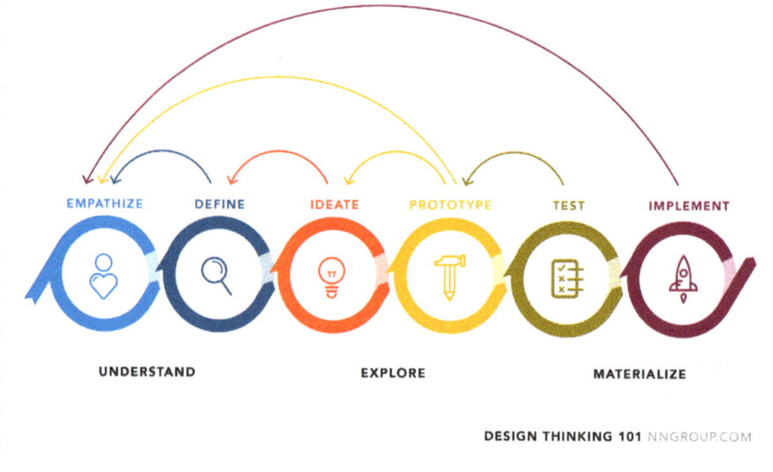

FIGURE 2.4
The design thinking process, visualized by the Nielsen Norman Group.

- **Double Diamond:** Formalized by the British Design Council, the Double Diamond framework emphasizes four common steps that teams follow when innovating and developing new products: discover, define, develop, deliver. The visualization in Figure 2.5 emphasizes the practice of divergent thinking and convergent thinking at each step of the process. This methodology reflects one of the primary responsibilities of a PM: knowing when to flare and continue problem exploration versus knowing when to narrow on a decision and advance to the next step.

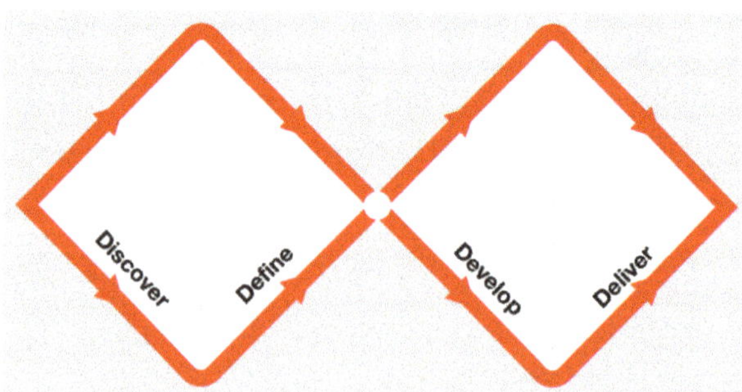

FIGURE 2.5
The Double Diamond framework, visualized by the British Design Council.

- **Jobs-to-Be-Done (JTBD):** Invented by Tony Ulwick and popularized by Clayton Christensen, this framework focuses on identifying the functional and emotional goals that customers want to accomplish in a specific context. Unlike traditional market research that groups customers based on demographic attributes, the JTBD framework segments customers based on shared needs and desired outcomes. By framing innovation around "jobs" that customers need to be done, teams can uncover unmet or adjacent needs that traditional segmentation might overlook. Figure 2.6 shows the five key elements of the JTBD framework from Jim Kalbach's *The Jobs to Be Done Playbook.*[2]

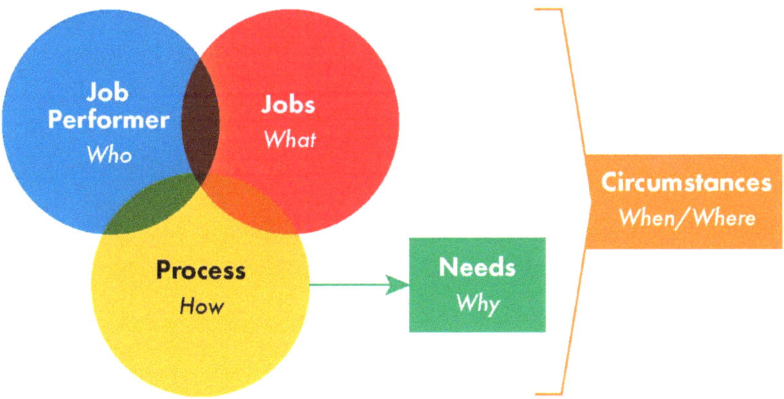

IMAGE COURTESY OF JIM KALBACH

FIGURE 2.6
The JTBD ecosystem, as depicted in *The Jobs to Be Done Playbook.*

- **Lean Startup:** Popularized in *The Lean Startup,*[3] this product development approach favors rapid iteration and customer feedback cycles to fuel product innovation. In his book, Eric Ries simplifies the process into three simple steps: build, measure, and learn. Through these steps, teams must clarify their hypotheses, build real-life experiments to test them, and gather instant feedback to validate or invalidate their hypothesis. This process continues until there is a decision to launch or pivot the product.

2 Jim Kalbach, *The Jobs to Be Done Playbook: Align Your Markets, Organization, and Strategy Around Customer Needs* (Two Waves, 2020).

3 Eric Ries, *The Lean Startup: How Today's Entrepreneurs Use Continuous Innovation to Create Radically Successful Businesses* (Crown Currency, 2011).

In truth, you can spend hours diving into each of these frameworks; they each have fan bases, case studies, and of course, critics! They are included here so that you can keep an open mind and understand some of the popular product design methodologies today. That said, new methodologies come and go as the industry evolves. Here are three pitfalls you can avoid when using these in your daily practice:

- **Don't forget about viability and feasibility.** Many popular frameworks emphasize customer-centric product design, but businesses need more than desirability signals to operate successful products.
- **Don't be a purist about these frameworks.** The perfect product design framework doesn't exist. Stay flexible. (And if your company is transitioning to a product operating model, don't be self-righteous about the "right" way to innovate or build products. Work with your business partners, not against them.)
- **Don't ignore the importance of domain expertise.** These frameworks are not a substitute for building industry expertise with your customers and stakeholders. First-hand user research is not the only way to learn. Follow industry experts, get a business mentor, and watch competitors. And don't forget to learn about industry constraints caused by compliance or regulatory requirements.

Software Development Methodologies

During product delivery, most teams follow strict software development life cycle (SDLC) steps to ensure that solutions meet quality standards and minimize risk for the business. The SDLC varies across companies (and sometimes varies among teams at the same company), but it generally follows these steps:

- **Planning:** Elicit and document project goals, objectives, and requirements. Estimate costs or resources needed and begin to document a high-level timeline with milestones.
- **Analysis:** Assess technical requirements in detail, identify potential risks and dependencies, and revise requirements or initial timeline based on feasibility constraints.
- **Design:** Define the solution design, including architectural or system design, user interface and interaction patterns, and backend elements.
- **Implementation:** Using inputs from prior phases, start coding and build the solution.

- **Testing:** Using manual and automated solutions, test the solution to ensure that functionality and usability satisfy requirements. Identify and fix any bugs or unacceptable issues.
- **Deployment:** Deploy the code into production (the product version that customers use).
- **Maintenance:** Use feedback and analytics tools to identify, fix, and resolve issues in production and continue enhancing and performing ongoing maintenance of the product.

To manage each step, product and development teams use workflow tools like Jira, Trello, or Azure DevOps to load tasks, view progress, understand project velocity, build timelines, and track dependencies all in one location. As a PM, you must master the company's workflow tools to keep the work and team on track during product delivery. That said, you may have the opportunity to shape how your team accomplishes the SDLC. Table 2.2 includes common SDLC models that teams follow when building and launching solutions.

TABLE 2.2 POPULAR SOFTWARE DEVELOPMENT METHODOLOGIES

SDLC Model	Description	Used By
Agile	Iterative process that emphasizes continuous delivery and ongoing feedback to adjust quickly. Scrum, Kanban, and Scaled Agile Framework (SAFe) are popular techniques.	Most companies today in order to ensure continuous delivery, especially when optimizing or maintaining existing products.
Waterfall	Linear development process that compiles all requirements at the beginning and delivers work in sequential phases.	Regulated and risk-averse industries when there is minimal appetite for continuous enhancements.
Iterative	Sequential process where work is divided into smaller iterations and the team produces working software at the end of each iteration until all requirements are met.	Teams that deliver proofs of concept or early product versions. Helps minimize complexity and manage costs.
DevOps	Model that promotes continuous integration and deployment of software to reduce handoffs. Emphasizes tooling and automation to release software quickly and safely.	Teams that deliver updates to shared platforms or solutions. Typically combined with another method above.

Agile is the most popular SDLC methodology today, but teams use different techniques when adopting it. The three most popular Agile techniques are:

- **Scrum:** A structured approach that includes defined roles, team "ceremonies," delivery commitments, and "sprint" cycles to time-box the work and deliver software at the end of each sprint.
- **Kanban:** A less-structured approach that promotes continuous workflow on the team. In Kanban, teams visualize all work on a board, move it across workflow phases, and continuously pull new work from a backlog.
- **SAFe:** A structured approach meant to manage multiple Agile teams at once by emphasizing planning and review checkpoints at multiple levels of an organization to ensure that all teams are delivering quality software and minimizing risk.

The Role of Product Artifacts

Now that you've had a crash course in product theory and methodologies, let's talk brass tacks. Theory and frameworks alone don't make you successful. You can follow step-by-step methods and still produce a mess. The trick is using the theories and frameworks to form a strong point of view about your customer and solution and then make concrete decisions to guide the team's work. That's where *product artifacts* come in. They are the visible expression of your judgment, thinking, and decisions that set the direction for the team.

In many organizations, product managers own the product artifacts listed in Table 2.3. The PM builds them by synthesizing and distilling important assumptions, insights, goals, and decisions, typically in collaboration with their business, design, and engineering partners. Once the core team aligns, these documents become the tool for driving alignment across the broader organization. Whenever senior leaders or team members need clarity on *why, what, how, and when*, product artifacts represent the truth. You'll learn exactly how to craft these artifacts in Chapters 7–10, but there's loads of foundational work that must happen before you can create these. Chapters 3–6 covers the groundwork first.

TABLE 2.3 THE PRODUCT ARTIFACTS EVERY PM MUST CREATE

Artifact	Description
Product Vision	A short statement that articulates why your product exists, the North Star vision for your product, and the positive impact that your product will create. It should inspire the team.
Product Strategy	A document that articulates the strategic assumptions, high-level bets, or major investments you're making to accomplish the vision. It informs the roadmap.
Product Roadmap	A document that outlines the team's goals, work, and measurable milestones over a given timeframe for a given product or initiative. It ladders up to the strategy and vision.
Product Requirements	Detailed descriptions of what must be true for your product to accomplish the customer and business outcomes that your product serves. These are given to design and development teams.
Product Performance Updates	A document that updates critical stakeholders on the status of product initiatives and products in-market, using data.

Pulling It All Together

Now that you've got the foundation to think like a real product manager in the wild, it's time to apply it. You understand the skills and competencies to be effective, you've absorbed product theory, and you know which artifacts every PM should build.

Next, we're moving from theory to application. In upcoming chapters, you'll walk through step-by-step techniques for navigating the entire product life cycle—from defining problems and crafting a product strategy to launching and managing products in-market. With reusable templates, examples, and advice from industry experts, get ready to elevate your practice.

EXPERT ADVICE FROM PRODUCT LEADERS

THE PM X-FACTOR: WHAT SEPARATES GREAT PRODUCT MANAGERS FROM EVERYONE ELSE

There's no shortage of hard-working product managers out there. But a strong work ethic is table stakes in this career. What actually separates good product managers from great ones? I interviewed world-class product experts and seasoned leaders who have spent years coaching and developing product management talent. Here's how they spot exceptional PMs.

"The best PMs understand how to see the forest for the trees. They are detail-oriented but dynamic enough to grasp the bigger picture and the 'why' behind the work. They are experts at helping others understand how day-to-day execution fulfills the larger plan."

—Dan Olsen
Product management consultant, trainer,
and author of *The Lean Product Playbook*

"A passion for technology and the ability to climb the levels of abstraction. Remember what you saw, but understand the bigger meaning."

—Des Traynor
Co-founder and Chief Strategy Officer at Intercom

"Curiosity about your customer. Deep curiosity is absolutely required and an unwillingness to compromise on quality. Always fight for quality."

—Teresa Torres
Product expert and author of *Continuous Discovery Habits*

"There's one thing that separates great product managers from good ones: an incredible, intense degree of personal ownership. The second part is the ability to learn and be curious. Develop strong beliefs, loosely held and be willing to let things go when presented with better evidence."

—Vito Covalucci
Head of AI Insights Products at Amazon

"Those who prioritize perfection over experimentation, expect a straight line towards product-market fit, or get caught in analysis paralysis aren't cut out for this type of work. You need to be flexible and driven."

—Chetan Kandhari
Chief Innovation and Digital Officer at Nationwide

CHAPTER 3

Why: Defining the Problem and Opportunity

Sarah is a junior product manager at a B2C software company and just landed a new role on the customer acquisition product team. She's eager to impress her leaders and peers. During her first week on the job, Sarah's boss drops by to pitch an idea for a new A/B test, her tech partners inform her that the sales app requires major architectural upgrades, and she inherits a dense backlog from the former PM that includes tons of new functionality enhancements. By Friday afternoon, Sarah is drowning in solutions but starving for direction. With so many ideas to consider, where should she start?

The answer is simpler than she thinks. Great product thinkers start by asking "Why?"

Whether it's Day 1 or Day 1,000, the best PMs ask questions that establish *strategic context*—why certain opportunities are worth solving right now. To do this, they clarify company goals, key markets to capture, and internal beliefs about what it takes to win those markets. And they evaluate conditions that will affect the company's ability to accomplish its goals, including competition, market trends, customer insights, or emerging technologies—all before committing anything to the roadmap. This context informs team priorities, helps them build outcome-based roadmaps (rather than reactive wish lists), and enables them to assess why certain opportunities are smarter than others.

Asking why and building strategic context will feel like you're slowing down progress at first but remember that your job isn't to ship solutions; it's to ship the *right* solutions, in the right way. You can't do this without an informed point-of-view, so always start by establishing context *before* committing to solutions. Of course, you can't spend months gathering strategic context, so try the methods in this chapter to answer the most important questions first:

- What is the company's goal and corporate strategy right now and why?
- What market does the company want to capture and why?
- What's the strategic problem or opportunity to investigate and why is it important?
- Who is your competition and why do they matter?

PLAY 1

Leading a Business Strategy Workshop

What is the company's goal and corporate strategy right now? What market does the company want to capture? Why?

To be successful, every product manager must understand an important distinction: product strategy and business strategy are not the same thing. Business strategy is a distinct input that *informs* product strategy. In other words, business strategy usually comes first.

A clear *business strategy* outlines core company objectives, clarifies target markets, and describes the high-level activities or services the company wants to provide a given market over a given timeframe. A *product strategy* uses that information to determine a specific offering that will deliver product-market fit and accomplish business goals. A strong product strategy includes the specific needs to solve, the specific segments to serve, the value proposition on which the product competes, and solution details.

In companies that offer a few products or services, the relationship between business and product strategy is straightforward and easy to understand (see Figure 3.1). Large companies with several strategic goals often use multiple product strategies to capture different target markets and establish diverse income sources (see Figure 3.2). Both figures illustrate how company objectives flow into business strategy, which then shapes product strategy. Together, these elements form a cohesive framework that answers these questions: Why this direction, who are you serving, what are you offering, and how will you compete?

Without clear business objectives or target markets, product managers spin their wheels creating products that suffer from inadequate resources, stalled launches, or failure to deliver value. If you're on a team exploring a new problem statement or solution, but no one understands why, you have a strategy gap. Don't let this happen to your work. When goals and business strategy aren't clear, someone needs to drive the conversion—and that someone is you. Here's a method to get started.

Company	Sally's Suds-N-Stuff Our mission is to keep dogs clean, healthy, and happy.	
Objective & Target Market	Grow our customer base by 10 million households over the next two years and increase share-of-wallet for the 59.8 million U.S. households that own dogs.	
Business Strategy	Connect dog owners, groomers, and veterinarians in all 50 states to capture 5% share in animal health and wellness market.	Scale physical and digital footprint in all 50 states to capture 10% share in pet supplies and retail market.
Product Strategy	Create a digital marketplace that connects groomers and veterinarians with dog owners for same-day appointments and instant advice for urgent issues.	Create premier omnichannel retail experiences that offer a diverse catalog of personalized lifestyle goods for pet parents.

FIGURE 3.1

While this fictional company has two distinct business and product strategies, both support one objective and target market.

Company	Google Our mission is to organize the world's information and make it universally accessible and useful.		
Objective & Target Market	*Strategic Objective #1* Increase revenue by capturing more share of U.S. advertisers' revenues from 26% to 40% in the next 3 years.		*Strategic Objective #2* Expand market share of business customers that rely on cloud computing and digital transformation services.
Business Strategy	Scale a global advertising ecosystem by deploying proprietary, world-class ad tech that helps advertisers, agencies, and publishers grow their business.	Scale ad reach by establishing our hardware and software as the essential infrastructure for personal growth, health, and success in the digital age.	Diversify revenue streams by providing mission-critical cloud platforms and cybersecurity solutions to power the next generation of business operations.
Product Strategy	Build a premier, AI-powered marketing analytics suite that helps advertisers personalize campaigns and optimize marketing spend based on value.	Create a unified AI intelligence layer across our product ecosystem to help users instantly discover reliable, useful information across primary touchpoints (i.e., Search, Maps, Chrome, Android, and YouTube).	Create a unified, security-first platform that integrates cloud-native tools and AI-optimized infrastructure to ensure operational continuity for the modern enterprise.

FIGURE 3.2

Because Google has multiple objectives and target markets to capture, the company requires distinct business and product strategies to support different goals and customers.

Use When

- The team lacks strategic context to adequately prioritize discovery or delivery activities.
- Internal stakeholders keep suggesting new "features" but can't articulate why.

Average Time

3–4 hours total, including 1 hour to prepare, 1.5 hours to collect information via workshop, and 1.5 hours to synthesize

Try This

1. **Identify key sponsors and schedule a workshop.**

   Schedule a workshop and invite leaders who are accountable for setting strategy, resourcing initiatives, and reporting results to executive leadership. Look for VPs or heads of functions like sales, operations, business development, product, design, or engineering. In some cases, you'll need a C-suite leader to attend. If you can't get commitment from a critical business leader to attend, ask for a delegate who can represent their strategic point of view. And always include your design, engineering, and business peers so they can build strategic context, too.

PRO TIP **STAKEHOLDER WORKSHOP VS. INTERVIEWS**

One-on-one interviews are a great technique when you're conducting a listening tour or gathering context from executives, but group workshops are better when you need clear decisions. Workshops allow leaders to spot misalignment faster and make trade-offs so you can gain clarity quickly, and they save you from triangulating across meetings.

2. **Host a workshop and collect input.**

   During the workshop, use a physical whiteboard or digital canvas to mirror the questions in Table 3.1. Set the stage by explaining workshop goals and then facilitate an open discussion to answer the questions, ensuring that every participant engages. As participants respond, fill in the canvas so everyone can see the collective output.

   Be assertive and ask follow-up questions when answers are vague or assumptive. If leaders offer superficial or generic goals, push them to be as *specific* as possible. When participants start pitching new ideas and solutions, ask follow-up questions focused on the core assumption behind it. For example, ask "What business outcome do you expect this idea to target?"

TABLE 3.1 BUSINESS STRATEGY WORKSHOP QUESTIONS

Topic	Workshop Questions
Objectives and Goals	What are the company's goals over the next 1, 3, 5 years and why? (Examples might be to improve profitability by X%, grow market share by X%, identify new revenue streams, etc.)
Target Markets	What are the primary markets we want to serve? How large are those markets, and how much share do we have today? How much market share do we desire?
Strategic Assumptions and Opportunities	What changes or new developments will likely influence how the market behaves or affect our businesses? What will it take to win these markets? (For example, new solutions, competitive pricing, stronger customer experiences, new distribution channels, etc.) What evidence supports these assumptions?
Competitive Differentiation	How do we want to be known among competitors? What unique capabilities will give us an advantage in this space?
Limiting Factors	What risks or external factors might prevent us from accomplishing our goals? What must we trade off to pursue this objective fully?
Metrics and Milestones	What metrics will tell us if we're winning or losing? What critical targets or milestones must we achieve to stay on track toward our goals?

3. **Prioritize strategic decisions during the workshop.**

   After discussing each category, have everyone vote on the top two to three items with the highest impact for the company, starting with the first section and continuing from there. For tie-breaking scenarios, ask the business or executive sponsor to make the final call. When you're done, review the decisions to ensure they make sense holistically. (Don't skip this step! These decisions ultimately form the compass you'll use to guide future work and manage trade-offs. If you don't narrow down now, you'll encounter conflict later.)

4. **Communicate these decisions.**

   After the workshop, send a message and share the information with your team, leaders, and stakeholders. Keep everyone aligned with clear communication like this:

   "Greetings. We recently held a strategy workshop with our executive sponsors for [Initiative Name]. We aligned on important elements of our business strategy, including core objectives, target markets, strategic assumptions, and key metrics. These will guide our product planning for the next year. You can view the strategic document [here]. Please reach out within the next week if you have any feedback. Thank you."

LEARN MORE

STRATEGY

The absolute best reading on strategy is Michael Porter's 1996 paper "What Is Strategy?" found in *Harvard Business Review*.

The best podcast on strategy can be found by listening to "Intercom on Product: The Intersection of Company and Product Strategy," www.intercom.com/blog/podcasts/intercom-on-product-ep07/

PLAY 2

Building Problem/Opportunity Statements to Find a Hunch

What's the strategic problem or opportunity to investigate and why is it important?

When PMs inherit an existing product, start an innovation project, or learn about a new company goal, they often encounter the "fuzzy middle." This is the moment when business goals and target markets are clear, but the path to achieving them isn't. When teams lack a clearly-defined problem or opportunity to explore, they start wasting weeks doing research that leads to debates over scope.

A well-crafted problem or opportunity statement can help you overcome this ambiguity and build momentum. These statements decompose broad business goals into specific, data-informed challenges. They articulate what the problem is, who experiences it, and why solving it matters. When written well, they become the foundation for research activities, brainstorming sessions, and strategic direction.

Defining strong problem statements is a fundamental product management skill that applies to every stage of the product life cycle. Every PM should know how to craft one. Here's how.

Use When

- You're taking over an existing product, and you're not sure which opportunities or pain points to solve first.
- The team has a clear business goal, but the discovery scope feels too broad and lacks boundaries.

Average Time

3–5 business days, depending on data availability

Try This

1. **Identify and document performance gaps.**

   For this step, examine prior research and existing data to discover performance gaps or missed opportunities. For a well-rounded perspective, ask experts, review multiple data sources, and collaborate with your team to complete the following:

   - **Evaluate current business performance.** Of the following stages, where is the company's weakest performance?
     - Generating consideration from prospects and leads
     - Converting leads into customers
     - Convincing customers to adopt and activate product usage
     - Establishing revenue from customers or monetizing the solution
     - Retaining customers and preventing churn
   - **Audit the customer experience.** Of the following journeys, where do customers experience the most friction?
     - Sales journeys, including free trial enrollment, product selection, and payment tasks
     - Activation journeys, including account creation, initial setup, feature discovery, product customization
     - Servicing journeys, including routine tasks, new feature adoption, settings updates, customer support needs
     - Renewal journeys, including product ROI (return on investment) evaluation, competitor comparisons, plan and pricing reviews, payment updates

- **Identify structural problems.** Are foundational issues causing repeated performance issues?
  - Does the core value proposition still resonate with prospects?
  - Do we have evidence our solution has sustainable product-market fit?
  - Do we have evidence that our pricing model and distribution strategy still works?
  - Do we have evidence of product quality issues or broken functionality that impacts the customer experience? (Example: Usability insights, app stability or performance issues, error rates, unreasonable defect levels, critical customer support themes, etc.)
  - Why do customers stop using the product or cancel?
- **Evaluate market dynamics and emerging trends.**
  - What are the fundamental needs underpinning this market today?
  - How well are those needs served today? What needs are met with inadequate solutions or workarounds?
  - What missed opportunities or blind spots exist across the offerings, experiences, and business operating models in this market today?
  - What might drive this market toward a new solution?
  - What emerging trends or pressures are creating demand for stronger solutions?

PRO TIP **TRY THESE DATA SOURCES**

If you have time, interview customer support experts, observe product usage, and analyze product performance data to identify pain points for customers, the company, or both. For a well-rounded data set, try the following:

- Explore recent market research studies or customer survey data.
- Review existing business trends or product performance reports.
- Analyze user behavior and funnel metrics with clickstream tools.
- Watch session replay recordings of customers using your products.
- Analyze customer support trends.
- Perform competitive analysis.

2. **Size the impact and benefit of solving each discrete problem.**

   For each performance gap, customer pain point, and missed opportunity, evaluate the following:

   - How frequently does this issue occur?
   - What is the user impact of each pain point and missed opportunity?
   - What is the business impact of each pain point and missed opportunity?
   - If you could solve this problem and improve a key metric by a certain percentage, what measurable value would this create for the company?

3. **Draft problem/opportunity statements.**

   Take each problem or missed opportunity and distill them into statements that clearly define the issue, gauge the opportunity, and communicate the potential benefit of solving it. Use the format in the following bullets to craft a statement. Develop multiple problem statements for each observable issue.

   - **Problem statement for an existing customer segment:**

     We've observed [X issue], which is causing [Y impact] for [Z target user]. Our data suggests this problem is affecting [# of people] and costing our business [$ dollar value]. If we could solve [this metric] by [X%], we estimate [ABC in expected value]. How might we [build, change, or create solutions] for [target user] so that [insert desired outcome]?

   - **Opportunity statement for expanding to a new market:**

     The current state of [industry or domain] has focused mainly on [segments, pain points, etc.]. Existing products/services are not adequately addressing [this gap] in their [offerings, experience, or business operating model].

     Our data suggests this opportunity is affecting [# of people]. If we could solve [this gap] for [segment], as measured by changing [this metric] by [X%], we estimate we could realize [ABC in expected value]. How might we [build, change, or create solutions] for [target segment] so that [insert desired outcome]?

4. **Group statements into themes.**

   If possible, group the statements into themes. For example, are multiple problem statements indicating a larger problem in the sales funnel? Did you uncover loads of negative feedback about partner integrations that cause churn with end users? Or perhaps you noticed a single market segment isn't retaining well despite feature enhancements. If so, cluster them and give them a new label.

When smaller problems fit into a larger theme, you're likely to uncover the need for a new strategic priority that requires bigger investments (i.e., redesigning the sales funnel, upgrading a core capability, or standing up a new enterprise solution). These larger themes often form the basis of a new product strategy pillar or cross-cutting enterprise initiative that requires significant resources and focused attention.

5. **Prioritize problems/opportunities with key stakeholders.**

   Bring your key stakeholders together to discuss each statement and prioritize the ones to solve. (Invite the same people who were recommended in Play 1's, "Business Strategy Workshop.") During the discussion, introduce each problem statement, highlight common themes, and answer these questions:

   - Across the problems or opportunities we discovered, what would be most impactful in helping us accomplish our business objectives and win our target market?
   - Do any of these problems or opportunities require a new enterprise initiative or cross-cutting strategy across teams? If so, who will sponsor it?
   - Should we tackle these problems or opportunities in a specific sequence to drive maximum impact?
   - Should we complete any formal business analysis or additional research before selecting a problem to solve (i.e., impact analysis, cost-benefit analysis, etc.)?

   Once the group answers these questions, use the inputs to complete the "Business Strategy Workshop" exercise in Play 1. Then draft goals for specific research activities and begin aligning partners you'll need to begin product discovery activities, including user research, ideation, and concept testing.

LEARN MORE

PROBLEM STATEMENTS

For more inspiration on problem statements, check out the book *Lean UX: Applying Lean Principles to Improve User Experience* by Jeff Gothelf and Josh Seiden.

PLAY 3

Mapping Competitors

Who is the competition and why do they matter?

Great strategy doesn't just rely on being "better"—it articulates why your company or solution is unique and different from competitors and why customers will choose you. Of course, you can't know that without understanding the competitive landscape and your customers' alternatives—you need competitive research to inform product research, product design, and overall product strategy. If you struggled to answer the competitive differentiation questions in your business strategy workshop or you're just starting a new role, take time to assess the competition and build a competitor map.

Use When

- Your company has clear objectives, target markets, and strategic assumptions, but you're not sure if the company's strategic assumptions are unique enough to win.
- You're starting a new product assignment and want to understand how customers might view your company compared to the competition.

Average Time

4–8 hours

Try This

1. **Identify the competition and build a competitor map.**

   Build the scaffold of a competitor map (using Table 3.2 as an example) and begin listing your company's direct and indirect competitors in the map. Direct competitors are those who offer the same types of products/services to the same market. Indirect competitors offer different products or services that solve the same need within the market, albeit differently.

TABLE 3.2 COMPETITOR MAP TEMPLATE

Element	Key Questions	Competitor #1	Competitor #2
Value Proposition	Why do customers choose this product over others?		
Primary Tasks and Key Features	What user tasks does this product support? What key features does it offer?		
Unique Capabilities	What unique human or technical capabilities does this product leverage?		
Pricing and Revenue Model	How is this product priced? How does it generate revenue?		
Distribution Methods	How do customers find and convert to this product?		
Customer Sentiment and Reputation	What do customers and industry reports say about this product?		
How We Compare	How are we similar or different to this product? What must we change to be competitive?		

2. **Review their products or services.**

   After you've identified all direct and indirect competitors, research each one across the following dimensions and fill in the map:

   - What is the value proposition or core promise they make to customers?
   - What are the primary user tasks they support?
   - What are the primary features they promote most heavily?
   - What unique human or technical capabilities do they use to support these tasks or fulfill customer needs?
   - What is their pricing model for customers and how does it generate revenue?
   - How do they distribute their product to customers?

3. **Assess customer sentiment.**

   Next, it's time to assess what their customers say and think. Assessing customer sentiment can help you understand the table-stakes needs your solution must address or find gaps in what the competition offers. Gather data on why customers like or dislike the competition and add it to the template. You can do this by evaluating:

   - **Public reviews:** Analyze customer satisfaction ratings on app stores, social media, or review sites.
   - **Industry reports:** Review analyst trend reports from JD Power, Forrester, etc.
   - **Digital intelligence tools:** Assess web analytics for competitor websites to understand how customers find and convert to them. Tools like Similarweb and Ahrefs are great options to understand competitors' website traffic, keyword searches, audience insights, and conversion tactics.

4. **Use findings to inform your company's strategy.**

   In the final step, compare findings to your company's current strategy. Bring stakeholders together again, share results, and pose deeper questions about competitive differences. Ultimately, the team should answer these questions:

   - How are we similar to our competitors and how are we different?
   - What does our company do today that's better than our competition? What sources of power do we have in the market?
   - Why do customers choose us today?
   - Why do others choose the competition over us today?
   - Why will customers choose us in the future over competitors?
   - What must we change about our strategy to position ourselves for the future?

   This exercise can help you clarify what makes your company different and unique today or reveal the need for more differentiation. Use the insights to craft new assumptions and testable hypotheses for further validation during prototyping or concept tests.

LEARN MORE

DIFFERENTIATION TECHNIQUES

For additional insight on competitive differentiation, check out the "10 Types of Innovation" at www.deloittedigital.com/us/en/accelerators/ten-types.html.

Pulling It All Together

Developing strategic context is not a one-time event or linear checklist—it's a continuous loop. Once you understand the company's goals, target markets, strategic assumptions, and current opportunities, you'll have enough context to begin shaping specific discovery activities (like user research or concept testing goals).

Said differently, you don't need 100% certainty to move forward. Some details might feel fuzzy. That's normal! The goal of these methods is to give your team enough direction and a well-defined problem statement so you can focus discovery efforts, not perfect every single assumption. You'll fine-tune assumptions through the research and testing phase. We'll address these phases in Chapters 4, "Who: Defining Customer Needs," and 5, "What: Defining Solutions."

EXPERT ADVICE WITH DES TRAYNOR

STRATEGY STARTS WITH THE RIGHT QUESTIONS

As the co-founder and Chief Strategy Officer of Intercom, Des Traynor oversees all of R&D, including product, engineering, and design. We caught up to discuss how the best companies form strategy and what product managers can do when the strategy isn't clear. Here's what he had to say:

At Intercom, we believe that a well-defined company strategy is the biggest enabler of speed. No one accelerates to a stoplight, right? The team won't hustle if they know work gets stuck waiting in review cycles. That's why we spend one month tackling existential questions to determine our strategy and 11 months executing strategic choices every year.

Normally, we schedule annual off-sites to form our strategy, or we convene during a crisis when there's a new trigger that must be addressed. A trigger might be the emergence of a new technology, a new market entrant, a brand crisis, or a financial performance issue. In these meetings, we include leaders from all company departments and ask ourselves "Why are we here?" "How do we solve this trigger?" and "How much change can we really make?" From there, we make strategic decisions about markets we want to capture and company goals we want to achieve. These decisions create the holistic glue across all choices including product decisions, pricing decisions, and go-to-market plans.

Once the company's strategy is clear, we expect PMs to focus on two things: pace and adherence to the strategy. They move as fast as they can but stay on strategy. And they don't cause harm to the company by deviating from our goals.

Ironically, companies who don't do strategy well are often obsessed with speed, but they focus more on hitting dates, which requires constant escalations about features and decisions. If the company's strategy isn't clear, PMs should escalate the right questions to leadership. Be careful not to pull bad strategy out of a leader with questions like "Should we enable a workforce management tool or build an AI co-pilot feature?" Solution-specific questions like that often result in a la carte decisions with no cohesive narrative for the company. Instead, try questions like "Are we seeking new addressable markets, or are we trying to get more out of our current buyers? Are we trying to target individual verticals, or are we pursuing a horizontal product and building everything for everyone?" Ask the types of questions that, if answered well, will answer many questions.

CHAPTER 4

Who: Defining Customer Needs

Problem statements, design challenges, and company strategy slides usually have one thing in common—they have a strong hunch. The hunch is the core assumption solving X problem for Y market will result in Z outcome. Sounds clean and simple, but real life is never that straightforward. In many cases, these hunches start with a *vague* problem for a *vague* market. For example, you might receive a new product assignment that reads like this:

> We've observed that our digital shopping experience is clunky and confusing for shoppers, which is causing them to abandon our site before converting. Our data suggests this problem is affecting 20,000 daily visitors and costing our business $50,000+ in daily revenue. If we can convert these customers, we believe we could capture at least $1.5M in additional revenue per month. How might we improve the sales funnel for our prospective customers so that they convert to our products?

Clearly this problem statement tells us the sales funnel has issues, but it doesn't specify *who* the customers are, *what* is clunky about the experience, and *why* they're confused. As-is, it's not clear enough to host a brainstorming session. (And don't even think about asking tech to build a solution using a description like this!)

Stakeholders send PMs problem statements like this all the time. The best response? User research that defines *who* and *what* to solve for. User research removes guesswork and strengthens direction for the team. It produces insights needed to evolve ambiguous problem statements into specific ones, like this:

> We've observed three critical pain points in our digital shopping experience.
>
> 1. It takes too long to load for web customers.
> 2. It's missing helpful content on the product detail pages for mobile customers.
> 3. It doesn't offer flexible payment options for our small business customers.
>
> Our data suggests the first problem affects 10K monthly web users, the second problem affects 5k monthly mobile users, and the third problem affects 5K small-business owners across web and mobile who want to purchase our product but need flexibility. In aggregate, these issues cost our business $50,000+ in daily revenue. If we solve each issue, we could capture $1.5M

in additional revenue per month. How might we increase app speed, improve our product detail content, and offer payment flexibility so more customers convert?

See the difference? That's the power of user research! Before you can solve a problem well, you must understand the customer's perspective first. That means conducting user research before jumping to new solutions, especially when tackling new problems or opportunities. It doesn't have to be heavy-handed, but it should be thoughtful and structured. Focus on answering these questions first:

- What are my customers' functional needs, emotional needs, and biggest pain points?
- How do they solve these problems today? What workarounds do they have?
- What does success look like for customers? And what should we do about it?
- What is my research plan to learn about our customer?

User Research Methods 101

Throughout this book, I'll highlight the critical allies and partners you need to build successful solutions, and user researchers are one of those allies. User research is a specialized discipline with some of the smartest professionals you'll ever meet—a strong researcher should inspire and influence solution requirements and design choices. If you work consistently alongside professional user researchers, consider yourself lucky because many PMs don't have durable access to them. That said, every PM should understand a few basic research concepts to keep momentum high during product discovery, whether you're partnering alongside a professional researcher or working solo.

NOTE **TARGET CUSTOMER VS. TARGET USER**

I'm using the terms *target customer* and *target user* interchangeably, but if you support B2B or B2B2C business models, it's nuanced. In B2B models, businesses may be your target customers, while their employees are your target users. In B2B2C products, you'll have multiple customers and users. Be sure to account for customer and user segments during research.

Generative Research vs. Evaluative Research

Generative research produces critical insights about customer needs or "jobs to be done." It answers questions like "Which needs are most important to customers?" "How are they meeting those needs today?" and "How do they feel about current solutions?" It also uncovers deeper motivations and pain points, including functional or emotional constraints.

Teams should use generative research at the *beginning* of the product development life cycle to understand user needs, fuel persona development, and prepare for brainstorming sessions. Popular generative research methods include user interviews, open card sorting, and open-ended surveys.

Evaluative research answers how well the design of an actual *thing* (like a prototype, live version of your mobile app, or even wording on the home page) meets users' goals. Within evaluative research, there are two types of methods: formative methods and summative methods.

- Formative methods evaluate the product's *usability*. They answer questions like "What tasks do users have the most friction with in our experience?" or "What parts of the experience work well and what parts need improvement?"
- Summative research delivers insights about the product's *overall design* compared to competitors or earlier iterations. It answers questions like "Are we improving over time?" or "How does our product's interface perform compared to industry benchmarks?"

Teams should use evaluative research *later* in the product development life cycle once the team is ready to study the impact of tangible design decisions. Popular evaluative research methods include concept testing, usability or task completion testing, heuristic evaluation, close-ended surveys, and A/B testing.

Attitudinal Research vs. Behavioral Research

There's another key distinction across user research methods: attitudinal versus behavioral methods. Attitudinal research reveals what users *think or feel* about a solution, including their preferences, likes, dislikes, and why. Behavioral methods reveal what users *actually do* in a specific context with a specific solution. Both methods can produce qualitative and quantitative data. Figure 4.1 helps break down options.[1]

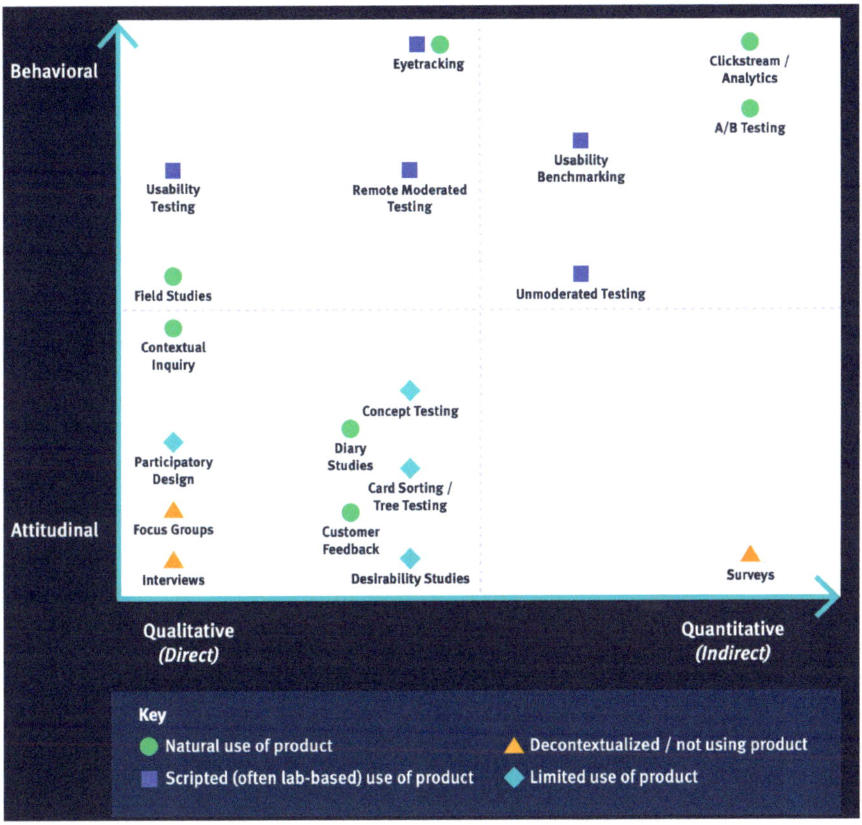

IMAGE COURTESY OF NIELSEN NORMAN GROUP

FIGURE 4.1
Use this popular research reference guide by Christian Rohrer when you want to determine the best research methods for your research goals. Graphic design by Kelley Gordon.

1 Christian Rohrer, "A Guide to Using User-Experience Research Methods," Nielsen Norman Group, July, 17, 2022, https://media.nngroup.com/media/articles/attachments/User_Research_Methods_A4-compressed.pdf

Popular Methods PMs Should Learn First

If you're just getting started with user research, familiarize yourself with the five methods listed below. These will come in handy when you don't have a dedicated researcher available to help and you're expected to tackle product discovery efforts on your own.

- **In-depth interviews:** One-on-one discussions with target users or customers to uncover motivations, beliefs, and thoughts about a particular topic.
- **Surveys:** A quantitative technique to assess participants' attitudes with closed-ended or open-ended questions.
- **Concept testing:** A technique where teams share solution ideas or prototypes with target users and gather feedback on the value proposition or solution requirements.
- **Usability Testing:** A technique used to evaluate how easily a user can interact with a product to accomplish their goals. Users are given a set of scenarios to complete and researchers assess users' effectiveness, efficiency, and satisfaction.
- **A/B testing:** Two or more variants of an experience are shown to users at random and statistical analysis measures the effect of each treatment on user behavior.

Fortunately, the user research community is vocal and produces tons of helpful templates and life hacks to navigate product discovery, so you don't need to start from scratch each time you need to conduct user research. If you're just beginning to learn about this topic, or want to deepen your knowledge, check out the books and thought leaders I've listed at the end of this section. Some of them even offer online courses, coaching sessions, and certifications.

And one more thing: Even if you partner with professional researchers, PMs should be deeply involved in shaping and reviewing research plans, observing research, and participating in synthesis to develop a deeper understanding of customers. Don't outsource your own learning!

LEARN MORE

TOP USER RESEARCH RESOURCES FOR PRODUCT BUILDERS

Just getting started with user research? Learn more with these resources.

- *Continuous Discovery Habits: Discover Products That Create Customer Value and Business Value* by Teresa Torres
- *Just Enough Research* by Erika Hall
- *The User Experience Team of One*, 2nd ed. by Leah Buley and Joe Natoli
- Nielsen Norman Group. For a list of research templates and cheat sheets, check out www.nngroup.com/articles/free-ux-templates/

PLAY 4

Building a Research Plan

What is my research plan to learn about our customer?

Understanding when to use different research methods is somewhat easy. Actually planning and doing the research? Not so easy. That's why the best-case scenario is to partner with a professional user researcher. Under ideal circumstances, they lead research planning, scoping, analysis, and synthesis activities while you and other teammates contribute, learn, and ride along. That said, you might find yourself without this level of support. When that happens, don't expect to become a professional researcher overnight (or ever, unless you switch careers and go all-in), but you can set yourself up for success by creating a simple research plan.

A research plan defines the goals, methods, timelines, and expected outcomes from user research activities. It provides structure for the team's next steps and increases the likelihood of producing actionable insights that guide solution development. Strong research plans can inform persona development, experience maps, and ultimately user requirements.

Research plans also help manage stakeholder expectations when the team is in learning mode. Unfortunately, some organizations are still obsessed with seeing a list of features and delivery dates to feel confident that they're solving the problems. If you're working for one of these companies, sharing a time-bound research plan will establish confidence that you're working quickly to identify a solution, de-risk incorrect assumptions, and avoid wasted efforts. But keep in mind, it's important to stick to the research timeline to maintain trust. You can build a research plan on your own or with members of your team in less than a day. Here's how to do it.

Use When

You're managing an initiative where details about the target user and their needs are fuzzy, and you don't have access to a professional researcher.

Average Time

2–4 hours, plus time to integrate feedback from stakeholders

Try This

1. **Define the research purpose and questions.**

   Block off an hour or so and write down exactly what you want from the research and why. Answer these questions:

   - What is the purpose of this research?
   - What decisions will this research inform?
   - What are the most critical questions we need to answer in this study?
   - What is out of scope for this research?

2. **Specify research participant details.**

   After defining the overall research goal, clarify exactly *who* you want to study and how you will locate them.

   - Who are the research participants we need to study? (Include characteristics like behaviors, goals, or familiarity with specific problems, tasks, or solutions.)
   - How many participants do we need in this study?
   - What requirements make a participant eligible or ineligible for this study?
   - How will we recruit participants?

3. **Identify the metrics you want to collect.**

   Early customer research focuses more on generating insights and less on measuring impact, but if you're using this guide to build a usability test, run a survey, or collect quantitative data, list the metrics you need to measure.

   - What attitudinal metrics do we want to collect?
   - What behavioral metrics do we want to collect?

4. **Select the research method you will use.**

   For this step, refer to Figure 4.1 and the popular user research methods listed earlier in this chapter to decide which approach is best, based on your objectives.

5. **Define the research outputs or deliverables.**

   Research without documentation goes nowhere. Creating research artifacts allows you to influence stakeholders and build a collective understanding of the customer. Therefore, every research plan should articulate the outputs from research. Common deliverables often include:

   - Research readout, including top insights, excerpts and recommendations
   - Customer personas
   - Customer needs analysis, including jobs-to-be-done maps or task analysis
   - Customer experience maps
   - Usability test results
   - Survey results

6. **Include a timeline of events.**

   Once you've nailed down the high-level research goals and method, build a realistic timeline of events. (Keyword here is *realistic.*) Most research takes at least one to two weeks, if not longer. Aim for around 2–5 weeks, but check with your leader to align on timing expectations and dependencies. If you need a buffer, plan for more time than you think is necessary. Then add these milestones to your timeline:

   - Design research.
   - Recruit users.
   - Conduct research.
   - Synthesize insights.
   - Share readout.

   After you've got it all figured out, compile this information in a one-page artifact that makes it easy for stakeholders to review and follow along. See Figure 4.2 for a sample template you can try out.

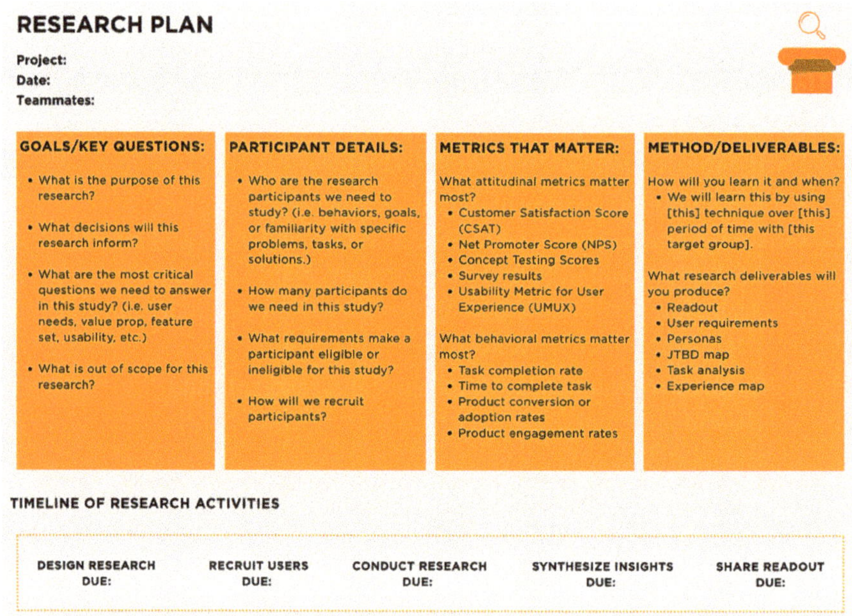

RESEARCH PLAN

Project:
Date:
Teammates:

GOALS/KEY QUESTIONS:
- What is the purpose of this research?
- What decisions will this research inform?
- What are the most critical questions we need to answer in this study? (i.e. user needs, value prop, feature set, usability, etc.)
- What is out of scope for this research?

PARTICIPANT DETAILS:
- Who are the research participants we need to study? (i.e. behaviors, goals, or familiarity with specific problems, tasks, or solutions.)
- How many participants do we need in this study?
- What requirements make a participant eligible or ineligible for this study?
- How will we recruit participants?

METRICS THAT MATTER:
What attitudinal metrics matter most?
- Customer Satisfaction Score (CSAT)
- Net Promoter Score (NPS)
- Concept Testing Scores
- Survey results
- Usability Metric for User Experience (UMUX)

What behavioral metrics matter most?
- Task completion rate
- Time to complete task
- Product conversion or adoption rates
- Product engagement rates

METHOD/DELIVERABLES:
How will you learn it and when?
- We will learn this by using [this] technique over [this] period of time with [this target group].

What research deliverables will you produce?
- Readout
- User requirements
- Personas
- JTBD map
- Task analysis
- Experience map

TIMELINE OF RESEARCH ACTIVITIES

DESIGN RESEARCH DUE:	RECRUIT USERS DUE:	CONDUCT RESEARCH DUE:	SYNTHESIZE INSIGHTS DUE:	SHARE READOUT DUE:

FIGURE 4.2
Keep the research plan to one page so partners can easily understand goals, questions, and timelines.

After you draft the research plan:

- Socialize it with your team and leader, integrate their feedback, and discuss trade-offs or risks that could set you back, such as staffing, recruitment issues, or competing demands.
- Start building the research guide and recruiting participants immediately. This can take longer than you expect. Herding cats isn't easy!
- Schedule the research and synthesis sessions in advance and include stakeholders on the invite. Insights are always much stronger when more people participate. Never do research or synthesis alone if you can avoid it.

PLAY 5

Eliciting Needs via Customer Interviews

What are my customers' functional needs, emotional needs, and biggest pain points? How do they solve these problems today? What workarounds do they have?

Remember that ambiguous problem statement at the beginning of the chapter? If you sent it to your design and engineering partners, you'd get blank stares (and maybe a side-eye). Your team expects you to represent the voice of the customer, know exactly what's wrong, and explain what success looks like, so making assumptions about the customer is a recipe for disaster. It doesn't matter how smart you are; if you are new to a team, industry, or product, your assumptions will most likely be wrong until you hear firsthand from customers.

That's why Teresa Torres, the leading continuous discovery expert, recommends user interviews as the first user research method every PM should master. This powerful technique can inform brand new solutions, improve existing products, validate your hunches, or lead to strategic pivots. Do yourself a favor and take Teresa's advice. Here's a method to help you design and conduct user interviews.

Use When

- Your company lacks deep customer knowledge from firsthand interactions.
- You're leading a brand-new initiative (or managing an in-market product), and you need to develop a clearer picture of user needs, goals, and pain points.

Average Time

- 4–6 hours to build your interview guide
- 10 hours to conduct 5–8 interviews
- 1–2 hours to compile notes and prepare for synthesis

Try This

1. **Build an interview guide.**

   The best interviews feel like natural conversations to participants while delivering valuable insights to interviewers. It's a tough balance to strike, which is why professional researchers always use interview guides to prepare. Interview guides outline the script you (or the researcher) follow to ensure consistency across interviews and accomplish research goals. They typically include an intro, open-ended questions, follow-up prompts to elicit specific answers, and a wrap-up section. (See Figure 4.3 for a sample outline and questions to include.) If you're stuck on what to ask, refer to your research plan; questions should tie directly to the goals and key questions in that plan.

USER INTERVIEW GUIDE

Opening
- Greeting, sign consent form, start recording

Interview Questions
- Tell me a little bit about yourself. Where do you live, what do you do for work, etc.?
- Tell me about the last time you experienced [this problem].
- What were you trying to accomplish?
- What triggered you to think about [something] and take action?
- What steps did you take to accomplish [something]?
- How did you get started?
- What knowledge, people, tools, or processes helped you accomplish your goal? How did you discover them?
- What was easy about doing [something]?
- What was hard about doing [something]?
- Did you have any workarounds to help you accomplish [something]?
- What did you avoid doing?
- What was the worst part about doing [something] or what do you wish was easier?
- How did you feel about the experience?

Follow-Up Prompts
- "Can you tell me more about doing [that task]?"
- "Why was that important to you?"
- "Can you describe the environment you were in when [this] was occurring?"

Wrap-up
- Thank them & tell participant what to expect next

FIGURE 4.3
Craft an interview guide, practice with a co-worker, and refine.

PRO TIP **TRY THE CRITICAL INCIDENT TECHNIQUE**

Jim Kalbach, author of *The Jobs to Be Done Playbook*, recommends using the *critical incident technique* during interviews to overcome generic answers from participants. In his book, Jim recommends these three steps during interviews to gather specific details about a key incident:

1. Ask the participant to recall a specific incident where something went wrong.
2. Ask them to describe their experience, including what was wrong and how they felt about it.
3. Ask them to describe their ideal state or what they wish had happened instead.

2. **Run a practice interview.**

   After you've drafted the interview guide, it's time for a dry run. Find a partner and practice asking questions by following the guide. If questions feel clunky, then rewrite them. If the sequence feels off, make adjustments. If you need to record or test audio tools, now's the time to practice that, too. Doing this will help real-life interviews go smoothly.

3. **Conduct interviews.**

   Once you've recruited participants, it's showtime. Keep interviews to 60 minutes or less and conduct them 1-on-1 to maintain participant comfort. For remote interviews, ask observers to turn off their cameras and take notes. When you're ready to begin, follow these tips:

   - Ask participants to sign an interview consent form (find examples of these online).
   - Start recording the interview (if using Zoom or similar tools, turn on the transcript functionality).
   - Introduce yourself and set context about the interview.
   - Follow the interview guide and ask probing questions when the participant isn't being specific enough.
   - Wrap up the interview and thank the participant for their time. Set expectations on compensation or what to expect later, if applicable.
   - After 1–2 interviews, adjust the interview guide if you notice you're missing important questions.
   - Avoid synthesizing until the end of all the interviews.

PRO TIP **OBSERVE CUSTOMERS IN CONTEXT**

If you get a chance to observe the customer performing a task in real life, seize the moment. Interviews are a great method to build understanding, but some people struggle to remember specific details that could influence the design of your solution. For example, if your company is trying to sell car insurance to customers while they are at the dealer, observe the purchase experience in real life. Doing so will yield richer insights.

4. **Save interview documentation in a shareable location.**

   Even with today's amazing tools and file-sharing systems, I *still* see people sending hard files via email or putting critical documents in their personal drive. You can't build a collective understanding or customer intelligence practice by hoarding information or making it hard to find, so be a bridge-builder—put your product documentation in one location that everyone can access. Once you're done with interviews, upload the following items to one location and send out a link to interested stakeholders:

   - Recordings of interviews
   - Transcripts from interviews
   - Interview notes
   - Synthesis artifacts
   - Anything else that could help new people onboard to the project

5. **Schedule synthesis.**

   Synthesis is essential. If you haven't already done so, schedule synthesis meetings within days of conducting interviews. (Ideally, you aren't doing this solo.) The next section covers synthesis practices you can try.

Synthesis 101

The whole point of user research is to produce *insights* that lead to *decisions* about the solution. In plain speak, if the research doesn't lead to a "so what?" decision, then it's wasted. (Don't expect users to make these tough decisions for you during research, that's what you're getting paid to do!) That said, shifting from *data to findings, findings to insights,* and *insights to action* isn't straightforward, and some people don't understand the distinction between these terms.

To clarify their meaning, let's review Nielsen Norman Group's definitions.[2]

- **Data:** Refers to an unanalyzed collection of observations about users that may include transcripts, notes, metrics, or survey output.
- **Findings:** Describes patterns in collected data or summaries across multiple data points.
- **Insights:** Focuses on explanations of opportunities, based on other user research and business context.

To put this in context, once you wrap up user interviews (or any user research activity), all you've got is raw data that doesn't mean anything until you complete analysis and synthesis. Analysis occurs by breaking down data to understand all its unique parts. During synthesis, you make sense of the data by drawing broader conclusions across several data sets and forming actionable *insights*.

For example, if you're reviewing interview transcripts to identify what you heard, how often you heard it, and what the patterns are, you're doing analysis and creating *findings*. But findings alone don't lead to aha moments—you still need to synthesize them and generate insights. Synthesis leads to a "so what?" recommendation that should be contextual to your business or product goals. In other words, it should inform product requirements, roadmap decisions, and user experience choices. And yes, ideally your research partner leads synthesis activities, but if you're working without one, here are three popular synthesis techniques you can try with the team after user research.

Technique 1: Qualitative Coding via Spreadsheets

Coding qualitative insights in a spreadsheet is old-school, but it's a comprehensive method. (*Coding* in this sense means categorizing the raw quotes, interview excerpts and observations in a structured format, not coding like a software engineer.) This technique forces the team to review all the user research data and think critically about its meaning. There are several ways to do this, but here's a simple approach:

1. Build a spreadsheet that lists research questions across one axis and interview participant names down the other. Pull the research questions directly from your Research Plan. (See Figure 4.4 for a template.)

2 Sara Paul, "Data vs. Findings vs. Insights: The Differences Explained," Nielsen Norman Group, April 23, 2023, www.nngroup.com/articles/data-findings-insights-differences/

2. With the team, review each interview transcript closely and pull out excerpts, stories, or raw data that correspond to your research questions. Then add the data in the correct cell. Rinse and repeat for each question and interview.
3. Review the raw data in each column, identify common patterns or themes, and discuss what it means about the target customers overall. Include any observations or findings that surprised you or invalidated a core assumption that the team had prior to research.
4. At the bottom of each column, add an insight or "so what?" statement that articulates what the findings mean for user requirements or solution insights.

	Research Question 1	Research Question 2	Research Question 3	Research Question 4	Research Question 5
Interview Raw Data					
Interview 1					
Interview 2					
Interview 3					
Interview Findings					
What patterns exist across interviews?					
What surprised us?					
Critical Insights					
Insight 1					
Insight 2					
Insight 3					
So, what does this mean for our solution or initiative?					

FIGURE 4.4
Recreate this template to code user research interviews using a simple spreadsheet.

PRO TIP USE AI TO SUPPLEMENT, NOT SUBSTITUTE

In today's age, PMs have the luxury of submitting transcripts into AI tools that can summarize findings instantly. Yes, this is an amazing way to supplement synthesis, but treat this as a supplement, not a substitute, to insight generation.

Technique 2: The User Interview Snapshot and Opportunity Solution Tree

Product discovery expert Teresa Torres recommends a two-step synthesis process that leverages two artifacts she created: the Interview Snapshot and the Opportunity Solution Tree. Here's a simple way to leverage this technique.

1. Using the Interview Snapshot in Figure 4.5, document what you heard in a *single* user interview. Fill in the experience map section by documenting the participant's story from *their* perspective, focusing on the critical moments. Complete this artifact for each interview.

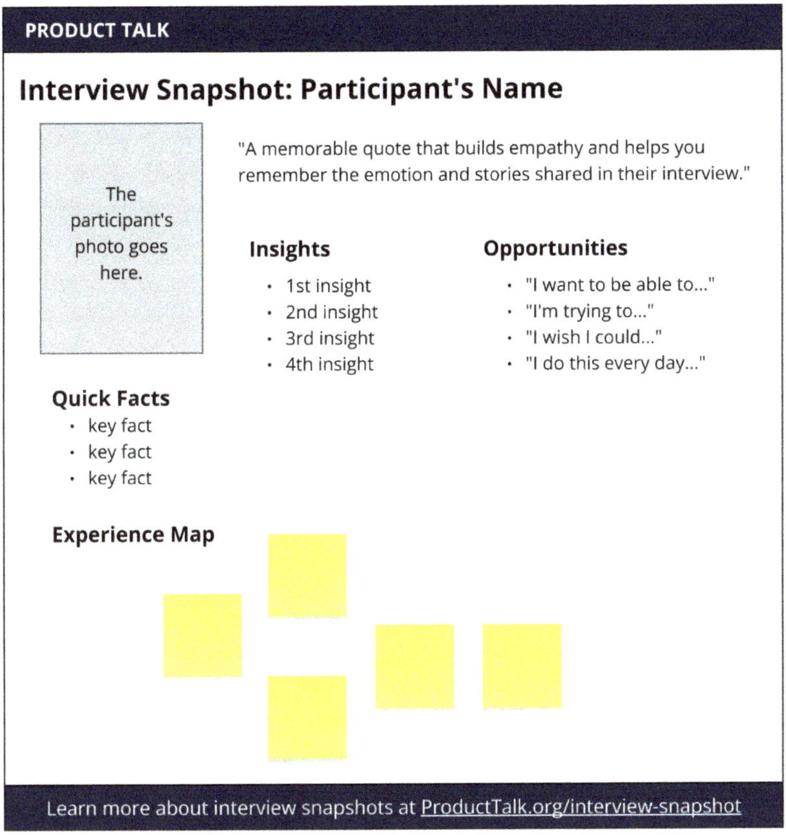

IMAGE COURTESY OF TERESA TORRES

FIGURE 4.5
Fill out an Interview Snapshot for every interview to remind yourself that no two users are exactly alike.

2. Start identifying critical *opportunities* from the experience map or other sections *across* your interviews. Opportunities could be unmet needs, memorable pain points, or customer desires. They should be meaningful enough that, if your company addresses them properly, a solution could lead to a significant customer or business value. Once you identify them, document them and move to the next step.
3. Start filling in the Opportunity Solution Tree, using the template in Figure 4.6. List the desired outcome or business goal at the top of the tree and add critical opportunities you identified in the last step. When you see multiple opportunities that could accomplish a similar goal, list them as sub-opportunities.
4. If the team has strong hypotheses about solutions, list those underneath the corresponding opportunities and start documenting experimental ideas to validate/invalidate assumptions about the hypothetical solutions. Repeat this process until you've mapped out all critical opportunities.

IMAGE COURTESY OF TERESA TORRES

FIGURE 4.6
Use the Opportunity Solution Tree to plot out unique opportunity areas and key assumptions when generating a new product strategy.

5. Once you've filled out the Opportunity Solution Tree, prioritize opportunities for future testing. Prioritize based on strategic fit, potential impact or expected upside, or importance to your target customer. Then add the test ideas to the team's backlog for future design research. Remember at this point, you are not picking solutions—you're prioritizing *opportunities* (i.e., pain points to solve, unmet needs to address, a suboptimized journey, etc.).

Technique 3: Jobs-to-Be-Done Mapping

Jobs-to-Be-Done (JTBD) is a popular framework used by companies around the world to identify customer needs and use consistent language about these needs across an organization. Several pioneers of this framework have their own twists on how to implement it, but I suggest using Jim Kalbach's approach from *The Jobs to Be Done Playbook*. Before attempting to use JTBD, you should read Jim's book first and study how others apply this method. That said, here's a basic introduction to the common definitions used in JTBD (shown in Table 4.1).

TABLE 4.1 JTBD DEFINITIONS

JTBD element	Definition	Example
Job Performer	The person who performs the job or task.	A first-time car buyer.
Functional Jobs	User or customer goals. These live at different hierarchies, and a "main job" or primary goal can include multiple sub-jobs along the journey.	Buy a new car.
Job Steps	Chronological steps a person takes to complete a job, including the beginning, middle, and end	Save for a car, get pre-approved for a car loan, browse car reviews, begin shopping for cars, select a car, finalize paperwork, and complete the purchasing process.
Needs and Success Criteria	Desired outcomes a job performer has when performing the job. These typically include a verb + noun + qualifier.	Decrease the time it takes to get preapproved for a car loan.
Circumstances	Details about when and where a job is performed.	Shopping for a car online from a smartphone app versus shopping at a physical car dealership.

Now that you have a basic understanding of the elements used in JTBD, here's how you can apply this technique after conducting user interviews or user research:

1. **Build a spreadsheet to structure user interview data.** Use Figure 4.7 for inspiration or adjust based on what works for your team. Begin filling in details from each interview with quotes and observations that answer the following questions:
   - What steps did the user take to accomplish the job or goal?
   - What were the user's needs or success criteria when accomplishing the job?
   - What pain points or friction occurred when accomplishing the job?
   - When and where did steps to the job occur?

	Steps to Complete Job	User Needs & Success Criteria	Pain Points	Circumstances
Interview #1 Raw Data				
Quotes or notes				
Quotes or notes				
Quotes or notes				
Quotes or notes				
Interview #2 Raw Data				
Quotes or notes				
Quotes or notes				
Quotes or notes				
Quotes or notes				

FIGURE 4.7
Spreadsheets are a simple way to structure user research insights into JTBD elements.

2. **Using the spreadsheet and interview data, build a job map or visual representation of the sequence that users follow when completing a main job.** Use Figure 4.8 as inspiration. The job map should reflect the beginning, middle, and end of the job, as well as the high-level stages of the journey along with the specific steps in each stage. The job map should also include useful details about the user's needs, success criteria, or critical pain

points under the relevant job step. This keeps the team grounded on needs versus functionality.

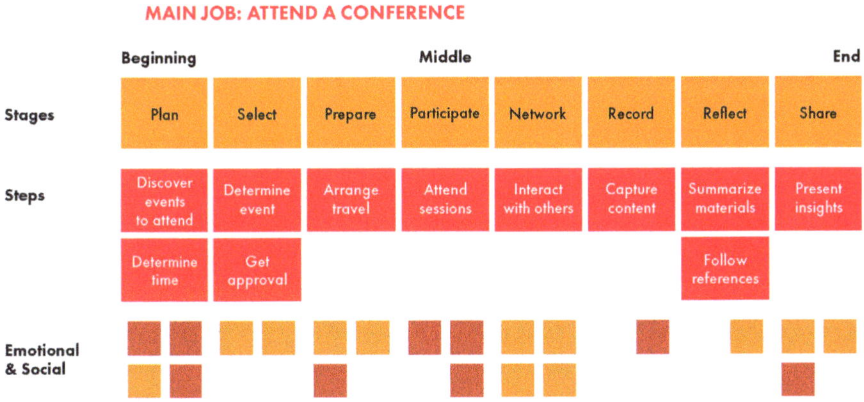

FIGURE 4.8

Job maps, like this one from *The Jobs to Be Done Playbook*, visualize the entire journey from the user's perspective.

3. **Review the job map and interview notes with your team and discuss where the biggest opportunities exist.** Ask yourself these questions to find meaning in the data and document what the team believes:
   - Is there a more efficient way to perform the job?
   - Where do people struggle the most to get the job done?
   - How might we eliminate stages or steps along the way so the customer doesn't have to take on as much work?
   - How might the job be carried out in the future, given current trends?
   - Are there related jobs we should address or fold into this job?
   - What could attract users toward a new or different way of completing the job?
   - What habits or anxieties might keep a user from trying a new solution or way of working?
4. **Based on the team discussion, identify the biggest opportunities or jobs to tackle.** Consider prioritizing by user importance, user satisfaction with current state solutions, or frequency of the job occurring.

LEARN MORE

SYNTHESIS

These three methods offer a starting point, not a comprehensive toolkit. Every PM should dive deeper into these techniques and explore additional synthesis methods like personas, empathy maps, and affinity diagrams. For additional reading, check out these resources.

- ***The Jobs to Be Done Playbook*** **by Jim Kalbach**
- **www.interaction-design.org/literature/article/methods-to-help-you-define-synthesise-and-make-sense-in-your-research**
- **www.nngroup.com/articles/thematic-analysis/**

PLAY 6

Sharing Research Readouts

What did we learn about customers and what should we do about it?

After clarifying the business strategy, conducting user research, and synthesizing findings, most PMs have a solid handle on what the business needs, what customers need, and which opportunities can bridge the gap. At this point, it's common for teams to revise their original problem or opportunity statement using new insights. But before you run off into concept testing, prioritize a pulse check with stakeholders, share research insights, and outline the next steps.

Research readouts keep partners aligned, informed, and supportive of your work. (Even teams with rapid experimentation cultures should perform this step.) And a word of advice here—whoever led the research should participate in the readout. In some cases, the PM and researcher should co-lead parts of the readout together, especially when translating findings into roadmap decisions.

Use When

User research and synthesis are complete, and the team is ready to share insights and next steps with key stakeholders.

Average Time

- 4–8 hours to build the readout
- 1 hour to host the readout

Try This

1. **Compile research insights into a brief presentation.**

   The best readouts include details about the study, insights about user needs and requirements, and specific, actionable recommendations for next steps. Most research readouts fall short on that last part and end with generic findings. (Don't let this happen to you!) You can avoid this by following this exact outline when building the readout:

   - **Slide 1: Executive Summary**
     - List the *most important* research findings, critical insights about the problem space or future solution, and recommended next steps and timelines.
     - List critical dependencies that will prevent you from moving forward (if any).
   - **Slide 2: Study Details**
     - Briefly list the target segment, how many customers you studied, and what method you used. Reuse this directly from your research plan.
   - **Slides 3–6: Key Findings**
     - Findings about the target customer (i.e., attributes, behaviors, preferences)
     - Findings about target customers' needs (i.e., importance, frequency, satisfaction)
     - Findings about current pain points and workarounds
     - Findings about solution requirements (i.e., user success criteria, including must-have and nice-to-have solution requirements)
   - **Slide 7: Surprises and Incorrect Assumptions**
     - What surprised the team?
     - What initial assumptions about the problem or users were wrong?
   - **Slide 8: Revised Problem Statement**
     - Include a revised problem statement (if applicable)

- **Slide 9: Insights for Our Product Strategy**
  - We believe the most important customer segment to solve for is [this] and why.
  - We believe the most important needs to solve are [these] and why.
  - We hypothesize [these value propositions or solutions] could address these needs, and we know we're right when we see [this metric or signal].
  - We believe the biggest risk right now is [this].
- **Slide 10: Recommended Next Steps**
  - What are the team's immediate next steps and timelines?
  - How will you validate/invalidate the team's hypotheses? (Example: concept testing, A/B test, quant studies, etc.)

PRO TIP **ALIGN INSIGHTS WITH BUSINESS GOALS**

To keep your audience engaged, don't bury the lede or overexplain. The best research readouts start with an executive summary, connect insights to business goals, and summarize the research in 10 slides or less. (You can always put more information in the appendix.)

2. **Present research at share-out.**

   By now, you know who the key stakeholders are, so schedule a readout and invite them to it. Send the readout in advance. (Some people prefer to read and digest content early). During the actual readout, try out this format:

   - Spend the first half presenting research insights. Leave time for discussion.
   - In the second half, facilitate Q&A.
   - Before concluding, ask project sponsors this question: "Based on what we know now, do you believe this problem/opportunity still aligns with our company strategy?" (This helps to secure ongoing buy-in and avoid surprises later.)

3. **Document agreements and update the roadmap or iteration plan for the team.**

   Finalize and publish guidance and agreements from the meeting. If you asked sponsors for resources, publish who's helping and when they owe a response. If needed, send follow-ups to ensure accountability.

Pulling It All Together

Mastering user research, synthesis, and readouts is an ongoing practice, so don't be afraid to roll up your sleeves and give it a whirl. Some research is much better than skipping research altogether, so design a plan, talk to customers, and share insights. And if you get stuck, don't re-invent the wheel—seek out articles, templates, and guides online for help. Plenty of PMs start by winging it and eventually become skilled at *just enough* user research.

EXPERT ADVICE WITH TERESA TORRES

THE #1 RESEARCH METHOD EVERY PRODUCT MANAGER SHOULD LEARN

Teresa Torres is a product discovery expert, coach, and the author of Continuous Discovery Habits: Discover Products That Create Customer Value and Business Value. *Here's the must-know advice she gives to PMs conducting user research:*

If there's one discovery method that PMs should master, it's customer interviews. During interviews, I strongly recommend story-based interviewing where we ask customers to share stories about their experiences with the problem the product aims to solve, rather than focusing solely on the product itself. This style of interview helps you keep the customer at the center of the story versus focusing too much on the solution itself.

I'll give an example. I did a lot of work with a marketing automation company when they were interviewing marketers to evaluate new solutions. They could have instructed the marketer "Tell me about the last time you used our product." That type of question keeps the conversation really steeped in the product and can result in limited findings from user interviews. Instead, they asked "Tell me about the last marketing campaign you ran."

By slightly adjusting how they asked the question, the interviewer opened up a much broader conversation about the end-to-end user journey. The product still showed up in that story, but the details about the customer offered a broader picture about their needs and opportunities just from asking the question differently.

After interviews, there are two levels of synthesis. The first is "How do I synthesize what I heard from this one customer conversation? And then the second is "How do I synthesize what I'm learning across my customer conversations?" That's why I recommend using both the Interview Snapshot and the Opportunity Solution Tree. By mapping out stories from the interviews, you can uncover new opportunities and insights to add to the Opportunity Solution Tree.

Once you've selected an opportunity to explore, the biggest mistake I see teams make after this is starting with one idea. Even if you design a good prototype with strong research questions, the feedback is always ambiguous because the user doesn't have anything to compare it against. That's why I encourage teams to work with at least three ideas. This allows teams to get feedback based on relative comparisons.

CHAPTER 5

What: Defining Solutions

Product management is a balancing act. You're expected to represent the voice of the user, while also navigating business realities. You obsess over ways to increase speed to market, while reminding leaders that quality products aren't born from deadlines. You put in countless hours to become an expert, just to realize how little you really know. After a day of nonstop meetings and context-switching, your brain feels like jelly. (If you don't have peaceful hobbies outside of this job, I strongly recommend you find some. I garden to stay sane.)

Yes, the life of a PM feels like a tightrope walk, but here's the honest truth: PMs are always slightly defining the problem and slightly defining the solution even after they've officially moved on from problem definition and ideation. That's because you're constantly learning what works and what doesn't by doing.

So, how do you increase the chances you're solving the right problem, the right way? You use methods that provide directional data to avoid strategic mistakes and plan for multiple rounds of iteration. This is where the trio of ideation, prototyping, and testing methods can help.

As I mentioned in Chapter 2, "The Effective Product Manager's Toolkit," the elements of product-market fit are like building blocks, and PMs need to align these elements together into a logical structure. If you don't have a solid grasp of the company objectives, target users, and user needs to serve, then go back and reread earlier chapters. But if the company goal, problem to solve, and customer needs are clear, it's time to actually do something about it.

When defining solutions, these are the questions you need to answer:

- What solutions solve the target customer's needs?
- Why would a customer convert to our solution?
- What internal requirements does the company have before launching a solution?
- What is our test plan to validate/invalidate early ideas?

User Experience 101

To create desirable and useful solutions, every product manager should have a basic understanding of user experience, design principles, and partnering with design teams. As the PM, you are the bridge across many functions. The best PMs learn each domain enough to build strong partnerships and elicit the best outcomes from each function on behalf of the company and customer. Consider this next section a mini-orientation in product design.

User Experience Defined

A product's user experience represents "all aspects of the end-user's interaction with the company, its services, and its products."[1] It encompasses the end-to-end user journey with your solution, including how users discover and interact with your products. It goes beyond what users see and includes what users think about your product, how they feel when using it, and how they interact with it.

Crafting a strong user experience requires excellence across many product design decisions, starting with what users don't see, including strategy, research, and testing, and ending with what users ultimately experience, including interface elements (see Figure 5.1). To do this well, design teams often include many experts that bridge design principles, psychology, engineering, and business expertise to create solutions that win customers and accomplish business goals.

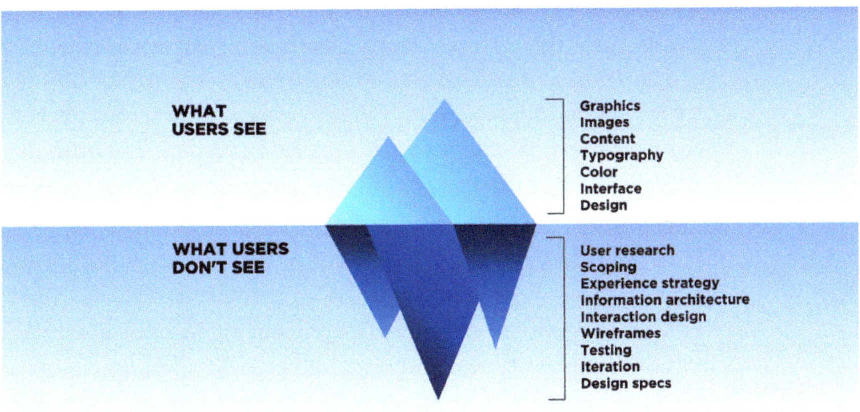

FIGURE 5.1
The iceberg analogy illustrates many layers and specialties in design that comprise the total user experience.

1 Don Norman and Jakob Nielsen, "The Definition of User Experience (UX)," Nielsen Norman Group, August 8, 1998, www.nngroup.com/articles/definition-user-experience/

Ultimately, the user experience represents the choices a product team makes, including what the team deemed important and where the team made shortcuts, concessions, or trade-offs during the product development process. Because the total experience can kill a brand or help it stand out among competitors, exceptional product managers consider user experience as part of the product's total value proposition.

Common UX Partners for PMs

User experience (UX) is a dynamic profession full of brilliant systems thinkers, visual designers, strategists, and doers. Most companies staff design teams in-house, although job titles vary across organizations. In general, PMs tend to work with these types of roles during the product development process:

- **User researcher:** These professionals are experts at designing, conducting, and synthesizing research to inform product decisions. They leverage quantitative and qualitative research methods to elicit critical insights about user needs and requirements. They also communicate the results via research readouts and assessments. Many companies have in-house research teams with job titles like UX researcher, user researcher, design strategist, or market researcher. Other companies partner with design agencies to conduct research. If your company has access to user researchers, figure out what methods and tools they use to accelerate product discovery activities.
- **UX designer:** These experts often work across the entire product life cycle. In smaller companies or companies with limited design support, these roles typically do a bit of everything. Most UX designers conduct some generative and evaluative research, might design prototypes, and often develop detailed mock-ups, wireframes, and specs for engineering teams to follow. They can also establish the information architecture, interaction design, visual design, and style guide used in a solution. Some organizations refer to UX designers as UX/UI (user interface) designers or experience designers, but in general, the responsibility is the same: understand user needs and design usable, desirable, and accessible solutions.

  Product managers most frequently work with UX designers. Their well-rounded skill set makes them an essential teammate on durable teams (often referred to as *pods, squads,* or *sprint teams*) where the product manager, designer, and tech lead form a cross-functional leadership trio for multi-month initiatives. To be clear, designers don't work *for* you, they work *with* you—so treat your design partner like a co-equal. On high-performing teams, the product manager, design lead, and tech lead align on purpose, team operating model, and plans to deliver solutions in-market.

- **Content strategist:** These specialists use the transformative power of words and content to communicate your brand ethos, influence user behavior, and provide much-needed clarity to customers. Sometimes referred to as *UX writers*, they ensure that the language across your product is clear, usable, and effective at helping users complete tasks. In companies where content is the product, content strategists take on a much larger role, often determining the content management model, including what content is generated, the process used to create and ship it, and how content is maintained.

  PMs should engage content experts during the initial design phase of a product to determine the optimal language in navigation menus, error messages, customer email communications, and critical user flows such as onboarding or renewals. PMs can also request content audits and experimental ideas from content strategists to improve usability and product adoption metrics once the product launches.

Requesting Design Support

A company's operating and staffing models usually dictate how and when product managers engage with design partners. On some initiatives, the product manager may have a consistent design partner embedded in the day-to-day work. On other projects, PMs must request design support on an ad-hoc basis. If your company has access to user researchers and designers, figure out how they staff and support projects.

Many design teams ask PMs to complete a "UX Project Brief" (see Figure 5.2), in order to gather details about the initiative so they can prioritize design resources appropriately. Companies often staff fewer UX roles than product managers, so you should complete this immediately and be prepared to justify the importance of your initiative. Also, be sure to include the opportunity size using dollar values when you submit the brief. If you already completed the business strategy workshop and problem statements from Chapter 3,

"Why: Defining the Problem and Opportunity," filling out a UX Project Brief should be easy.

UX PROJECT BRIEF

Project Name: Sally's Suds-n-Stuff, website conversion optimization

Background: As we expand our digital footprint into more states, we must ensure our online booking process is optimized for first-time and returning customers who demand a quick, seamless booking experience. We want to generate at least 10,000 additional bookings a month which will generate $800,000 in additional monthly revenue.

Objectives & Success Criteria: Reduce the % of customers who abandon during the booking process; reduce the time it takes for a customer to book a grooming appointment.

Request Details & Scope of Work: We believe user testing, a content audit, and a heuristic evaluation of our booking app will uncover opportunities and insights that will feed our optimization roadmap.

Key Stakeholders: CEO, CPO, and Team Wagtails

Timeline & Key Dates: Final deliverables due by Q3

Deliverables: Research readout from user testing, content audit, and heuristic evaluation, including recommended actions/improvements

Dependencies: Ability to recruit users in our new states

FIGURE 5.2
Including the opportunity size in a UX Project Brief helps design leaders prioritize staffing decisions.

Usability Heuristics, Design Principles, and Design Systems

A great product strategy poorly executed won't impress your customers if they can't use the product. Remember, usability is a non-negotiable quality of successful products. Fortunately, teams can set themselves up for success by following common usability heuristics, establishing clear design principles, and embracing design systems.

Usability Heuristics

Usability heuristics, or best practices, offer common guidelines that help teams create solutions that customers can use. Whether you have a design partner or not, product managers should understand common usability heuristics. This knowledge will help you assess your own product, build prototypes for testing, conduct competitor teardowns, and use the right language when communicating ideas to design partners. Refer to Figure 5.3 for 10 heuristics popularized by early usability pioneers like Jakob Nielsen and Rolf Molich.

For the most part, heuristics are consistent across UX educators. There are a lot of graphics out there, but they are all similar flavors of the same 10 heuristics.

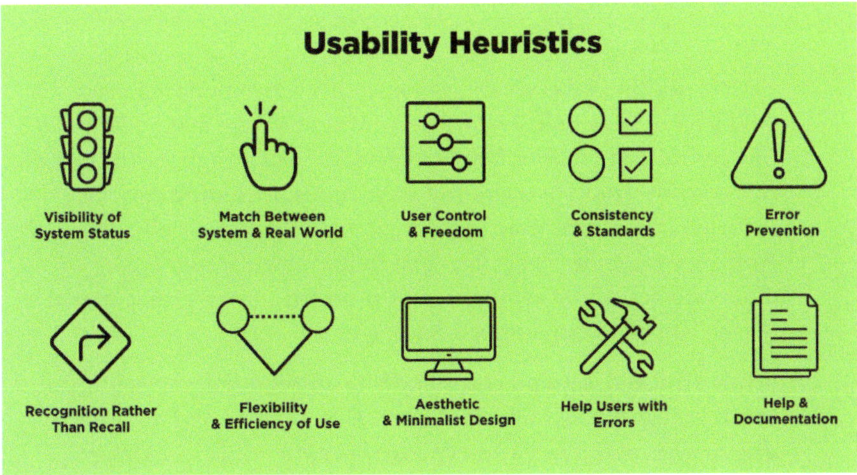

FIGURE 5.3

Heuristic analysis is a low-cost, high-impact technique to identify usability issues with your solution.

Design Principles

Design principles represent a company's values when it comes to product design. These principles help distinguish your product from competitors, communicate how the design should feel to users, and drive trade-off decisions when the team is evaluating different concepts to solve a problem. For example, a company's design principles could be the following:

- Effortless experiences that minimize the cognitive load on the end-user
- Efficient workflows that reduce the time it takes to complete a task
- Intelligent experiences that prioritize human-like guidance or expert advice for complex decisions
- Inclusive design that emphasizes accessible solutions for all
- Ethical experiences that respect the security and privacy rights of our users

If your company staffs an in-house design team, you may already have a set of design principles and guidelines to follow, so ask around. You can use them to ensure that the product stays on-brand with company expectations.

Ethical Design

Ethical design is everyone's responsibility when building products. And it's not just about avoiding harmful practices that make tasks like cancellation harder for customers. Ethical design means proactive consideration for users and their rights, including privacy, security, inclusivity, transparency, accessibility, and well-being. Take Apple for example. They don't just follow privacy laws—they've made privacy a core aspect of their products, proving that ethical design can be a competitive advantage.

While ethical design practices apply to all products and solutions, this is an especially important topic as automated and algorithm-driven systems power user experiences with minimal human oversight. In Chapter 6, "How: Defining the Approach," you'll find tools and questions to help teams define and assess ethical guardrails during product development, along with strategies to implement feedback loops and decision checkpoints. PMs can also take this one step further by embedding ethical design guidelines into product requirement documents.

Design Systems

Design systems are essential for scaling products and platforms to millions of users. Design systems provide reusable standards and code components to ensure that delivery teams maintain consistency within and across products. For example, color palettes, icons, page layouts, interaction patterns, code snippets, typography and more all live within a design system, typically located in digital libraries that designers and developers access.

As a PM, it's important to leverage and adhere to design systems to ensure usability and brand integrity. In addition to providing brand and experience consistency for end-users, they also help product teams ship faster and manage partner integrations. Product managers that own and maintain B2B or B2B2C platform solutions often publish design systems with their product documentation so partners understand exactly how to integrate it with their ecosystem. For example, companies like Spotify, Uber, Google, and Apple all publish design system guidelines for millions of engineers to follow, as shown in Figure 5.4.

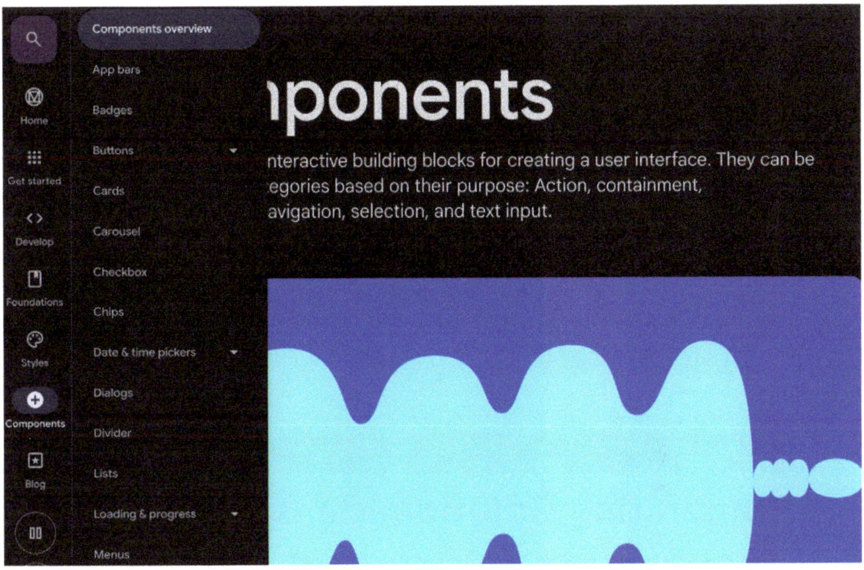

FIGURE 5.4
Google's design system is a well-documented guide that helps platform teams ensure consistency for users and the company's brand.

Continuing Ed

Even if you have the best design partners, accountability for a solution's success or failure often rests with the product manager. Do yourself a favor and stay curious about design practices. Thankfully, there's a vibrant online community of experts and thought leaders eager to help you succeed. And, if you're flying solo without an in-house design team, developing expertise in this field is essential. In that scenario, you generally have three choices:

- **Learn more about research and design on your own.** There are zillions of bootcamps, books, blogs, and online communities that can help you dive in. The absolute best authority on research and design is the Nielsen Norman Group.
- **Find a surrogate at your company who has a similar skill set to design.** People with titles like UX engineers, product marketers, or market researchers might be useful substitutes until you have official design support.

- **Ask for funding to hire contractor support.** (This usually takes too long and doesn't get approved if it's an unexpected expense, but it never hurts to ask.)

NOTE USEFUL DESIGN RESOURCES

The design community produces excellent literature, case studies, and resources to upskill practitioners around the world. Start learning more by exploring these books, thought leaders, and websites:

- *The Design of Everyday Things* by Don Norman
- *Don't Make Me Think* by Steven Krug
- *Rocket Surgery Made Easy: The Do-It-Yourself Guide to Finding and Fixing Usability Problems* by Steven Krug
- The Rosenfeld Media book library at **https://rosenfeldmedia.com/books/**

PLAY 7

Generating Solution Ideas

What solutions solve the target customer's needs?

After user research, teammates often jump straight to solutions. Sometimes, it's reasonable to fast-track obvious fixes and add a ticket to the development team's backlog. For example, if research indicates the navigation menu has a confusing label or you discovered a critical bug, engage an expert and resolve it quickly. You don't need to define, ideate, prototype, and test every change or minor adjustment.

However, when the right solution isn't clear and the cost of being wrong is high—such as a major redesign, platform migration, or product launch—you should leverage ideation, prototyping, and testing methods first. In these cases, you should explore multiple solutions before converging on a decision. On some teams, the design leader may schedule and facilitate an ideation session, but if you're unsure, or if you don't have a design partner, then you'll need to take the lead. (Someone has to!) Here's how to host one.

Use When

You've just wrapped up user research and need to generate testable solution ideas.

Average Time

- 2 hours to prepare
- 2.5 hours to host and synthesize ideas

Try This

1. **Select an ideation method.**

   The purpose of brainstorming or ideation sessions is to generate as many divergent ideas as possible before selecting which ones to test. Initially, the team should aim for quantity of ideas over quality of ideas. Like synthesis, this isn't an exact science, but there are plenty of useful methods to follow. Here are some of my favorite approaches from the Interaction Design Foundation.[2] Pick just one for your session.

   - **Brainstorming:** A technique that leverages the *collective* thinking of the group by engaging with each other, listening, and building on others' ideas.
   - **Braindumping:** A technique where individuals silently write down all of their ideas and share them one-by-one in a group setting without discussion. This method is intended to get everything out on the proverbial table before debating ideas.
   - **Brainwriting:** A technique where participants write ideas onto cards and then pass their idea cards on to the next person, moving those cards around the group in a circle as participants build on the ideas of others. Think of this as a "yes, and" exercise. It helps the team take a more collaborative approach and forces participants to consider others' ideas.
   - **Brainwalking:** A technique that emphasizes movement during the ideation session and gets participants moving around the room to review and build on others' ideas.

   During each technique, participants generate visual stimuli—typically sticky notes—to jot down and share their ideas. Keep in mind, others may prefer drawing or creating diagrams to communicate ideas. Either way, ensure that the environment is conducive for collaboration. If it's in-person, make sure that the

2 Rikke Friis Dam and Teo Yu Siang, "Learn How to Use the Best Ideation Methods: Brainstorming, Braindumping, Brainwriting, and Brainwalking," Interaction Design Foundation, last modified December 2024, www.interaction-design.org/literature/article/learn-how-to-use-the-best-ideation-methods-brainstorming-braindumping-brainwriting-and-brainwalking

room has useful materials (i.e., a white board, markers, sticky notes, etc.). If the session is remote, ensure that your participants have access and permissions to the digital tools you plan to use. (Tip: If you need to host remotely, the first three techniques can be done in-person or remotely with tools like Miro and FigJam.)

2. **Schedule and prep for the session.**

   After deciding on the format, it's time to start prep work. Skipping this step will lead to a poorly run session and a potential brand issue for you, especially if you're inviting influential stakeholders. (Alternatively, this could also be a chance to demonstrate strong communication and collaboration skills.) Become a master at facilitating these and set yourself up for success by doing this ahead of time:

   - **Schedule the session in advance and invite a diverse group to participate.** In addition to your everyday partners, it's important to invite key stakeholders and people with diverse perspectives. This could be senior leaders, your boss, customer service reps, or anyone with expertise in your problem space.
   - **Identify a co-host for the session.** It's hard to host and participate at the same time. It's also helpful to have an extra hand when "technical difficulties" occur or participants need support during the session. Make sure you're able to participate by finding a co-host and including them in prep work.
   - **Create an agenda.** Ideation sessions are usually 30–60 minutes and include multiple rounds of idea generation. Set a clear agenda with timeboxed increments to ensure that the group has enough time for silent work and group discussion.
   - **Draft prompts to structure the ideation rounds.** This is by far the most important part of prep. The group will need context and prompts to inspire and guide them. You can do this by drafting "How might we" statements based on the user needs, pain points, or underserved jobs-to-be-done you want to address. Here are a few examples:
     - *"How might we reduce the time it takes for patients to book a same-day appointment with their doctor?"*
     - *"How might we increase customer trust when submitting payment method information during the checkout process?"*
     - *"How might we maximize customer confidence when choosing their annual benefits package during open enrollment season?"*

    - *"How might we increase the speed at which we can integrate with third-party partners?"*
    - *"How might we decrease the percentage of new customers who call in to speak with a customer service representative?"*
  - **Document ground rules.** Groups don't always behave like you expect them to, so define ground rules for the session. Ground rules could be something like "refrain from judging or shutting others' ideas down," "no interrupting while others present ideas," or "stick to the prompts so we collect relevant ideas."
  - **Confirm the ideation environment is accessible and easy to use.** Some participants might join remotely while others might show up in person. Prepare an agenda that works for both groups. If you have remote attendees, test out tool permissions in advance.

3. **Host the session.**

   Ideation sessions are invigorating, collaborative sessions. Assuming prep is complete and your participants do show up, it's time for the fun to begin. During the session, follow these steps.

   - **Set expectations.** Share the agenda, ground rules, and purpose of the meeting. And stick to the plan.
   - **Offer context on the problem or opportunity.** Provide the original problem statement to the group and explain the opportunity you want to solve for the company and target customer. Then share user research findings, including the top user needs you discovered, critical pain points, and current workarounds. (This should be easy-peasy if you completed Play 6, "Sharing Research Readouts," in Chapter 4, "Who: Defining Customer Needs"? after user research.)
   - **Start ideation rounds.** Before each round, review the "How might we" prompts you want the group to follow. During the round, use a timer to stay on track.
   - **Review ideas, discuss, and iterate.** After each round, review and discuss the proposed ideas. Ask clarifying questions, if needed.

4. **Group ideas from the session.**

   After ideation, group similar ideas to identify common themes and distinct hypotheses. This clustering activity helps organize ideas that span different aspects of a solution. For example, some ideas might represent an overall value proposition, while others represent specific features or even improvements to the business model. To help with this, consider mapping ideas into these buckets:

- Distinct value propositions
- Discrete features or capabilities
- Unique design principles or brand principles
- Customer service changes
- New business models or pricing models
- Changes to profitability levers
- New ways to distribute or position the solution
- Business process changes

5. **Prioritize ideas for testing.**

   After grouping, select the ideas that you will prototype and test. Many teams use an impact-effort matrix to prioritize concepts, as shown in Figure 5.5. Using this matrix, plot ideas based on their effort (the level of effort it takes to deliver the solution in-market) and impact (the solution's assumed value for your company or customers). After you plot the ideas in this matrix, focus on delivering quick wins and big bets. Quick wins typically don't require testing because effort is low and impact is high. However, you should test ideas that fall within "big bets" before making a large investment in the solution.

FIGURE 5.5
Use the impact-effort matrix to plot hypotheses in the correct bucket after ideation.

Assumptions Mapping is an alternative technique you can use to prioritize hypotheses for testing. In this method, you'll plot hypotheses on a 2 x 2 matrix where the Y-axis represents how *important* the hypothesis could be to customers or company, and the X-axis represents how much *knowledge* or *evidence* you have about the hypothesis (see Figure 5.6). This method comes from David J. Bland's Assumptions Mapping technique.[3] When using this method, David recommends testing hypotheses that fall in the upper-right quadrant, known as "leap of faith" assumptions to minimize company risk while pursuing high-impact solutions.

IMAGE COURTESY OF DAVID J. BLAND

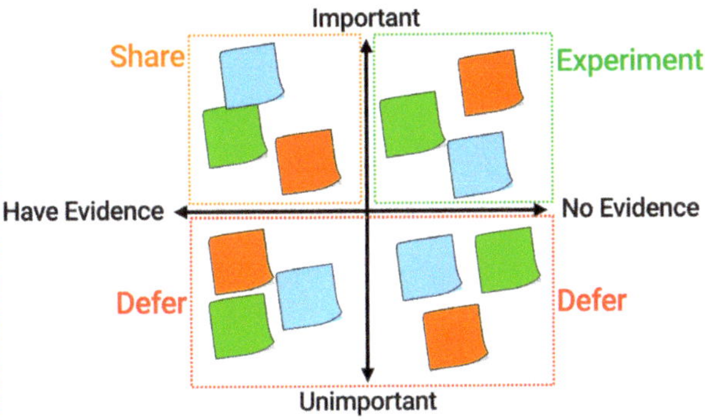

FIGURE 5.6
When using the Assumptions Mapping technique, test hypotheses that fall in the upper-right quadrant.

PRO TIP **TEST THREE IDEAS OR MORE**

Regardless of which prioritization exercise you follow, pick at least three options to test so that you have hypotheses to compare during testing. This will improve the quality of customer feedback and learnings you uncover during concept testing rounds.

3 David J. Bland, "Assumptions Mapping Worksheet," Precoil, 2025, www.precoil.com/assumptions-mapping

EXPERT ADVICE WITH VITO COVALUCCI

SUPERCHARGING IDEATION WITH AI TOOLS

Product discovery cycles are now faster than ever, thanks to AI tools. Vito Covalucci, head of AI Insights Products at Amazon, shared how their product teams use AI tools to reimagine and accelerate how they ideate and prototype solutions.

At Amazon, we're using AI tools to accelerate every stage of product development. Right now, we need discovery to be more fruitful and faster for PMs. Historically, ideation sessions have been full of sticky notes and the occasional drawing or diagram, but generative AI tools offer a new way to generate and communicate ideas and hypotheses during ideation. Like vibe-coding, we're encouraging PMs to practice videation, which means we want teams leveraging AI tools during ideation to instantly create prototypes and stimuli to share, discuss, and debate as a team.

Anyone can try this, but if everybody's using ChatGPT to run ideation sessions, you'll see very similar, nondeterministic answers. To generate unique, differentiated concepts, PMs must be able to write strong, nongeneric prompts and be willing to customize AI solutions to your company, instead of using off-the-shelf versions. That means giving the tools access to core insights that your competitors don't have, including past documentation, artifacts, experiments, interview transcripts, data, etc. All that history and data can lead to breakout moments, customized recommendations, and stronger solutions during ideation.

Keep in mind, AI-generated recommendations aren't a substitute for actual research and experiments. Don't start acting until you clearly define hypotheses and have strong data about those hypotheses. There aren't any true shortcuts to product growth or optimization. Deep research and testing are still critical.

PLAY 8

Testing Product Concepts

What solutions solve the target customer's needs? Why would a customer convert to our solution? What is our test plan to validate/invalidate early ideas?

Concept testing is an exciting moment in the product development journey. It's when you finally get to create something with all the user research you uncovered and assess how well it solves the problem statement that keeps nagging at you. Once again, you don't need to test everything. But when the cost of making a wrong choice could result in financial risk, brand risk, negative customer impact, or metrics that move in the wrong direction, then it's time to run a test.

There are plenty of ways to design experiments and learn what customers do when given a solution or why they do it. For example, product teams can do the following:

- Show simple sketches or paper-prototypes to users.
- Run a quantitative survey that asks users to rank unique value proposition phrases, images, or videos that describe a new solution.
- Ask users to create their own product by presenting a set of discrete value proposition statements and features on flash cards and then ask users to create their ideal product.
- Present a high-fidelity mock-up or clickable prototype that introduces a new website or app to users.
- Launch a fake-door test (a live landing page for a product that doesn't yet exist) and measure how many visitors sign up or demonstrate demand.
- Run a sequence of multiple experiments that use the different methods in bullets above and consolidate learnings across each to form a cohesive product strategy.

If you're testing enhancements to an existing product or platform, you may have better options available to you, including A/B testing or multivariate testing. (For details on that method, check out Chapter 10, "Managing Products In-Market," which includes techniques to optimize in-market solutions.) In this method, we'll explore the basic steps of concept testing to assess hypothetical value propositions and feature sets with users.

Use When

You have an early-stage product idea that requires additional validation before making a big investment to build it.

Average Time

5–10 days, including test design, building a prototype, running research, and analyzing results

Try This

1. **Clarify the goal of your test.**

   Before testing a product concept, ensure that you have a clear list of user needs or problem statements and a corresponding set of hypotheses to study. Then clarify exactly what you hope to learn in the concept test. Most concept tests aim to answer the research questions shown in Table 5.1.

TABLE 5.1 RESEARCH THEMES WHEN TESTING CONCEPTS

Theme	Research Question	What You Can Learn
Value Proposition	What is the primary benefit statement that resonates most with customers?	• Why customers might be willing to choose your solution ("reasons to believe") • What customers might be willing to pay • Why customers choose competitors instead
Functionality	What features would customers find desirable and useful?	• What functional requirements users have when solving a job-to-be-done • What features are table stakes, must-haves, and nice-to-haves in your initial product
Experience	What elements of a user experience would customers find desirable and useful?	• What customers feel and think when interacting with your solution • What design principles your solution should follow

A word of caution here—don't jam too many variables or hypotheses into one test. Once you start co-mingling hypotheses in an experiment, it's nearly impossible to understand how each hypothesis performed and why. If you're testing more than two elements in the table, break the test into multiple tests to improve your learnings.

2. **Determine key metrics.**

   Now it's time to apply rigor and determine what must be true to call your hypotheses a success or failure. At this point, you may have many hypotheses or just a handful. Either way, clearly document them in the same format and keep track of them in one location. Table 5.2 shows a simple template you can build to consistently frame test language across user needs, corresponding hypotheses, and metrics. (If you don't know which metrics to list, skip this step but come back and complete it after you select the research method. Metrics may depend on the method. For example, if you are running a survey, you could measure success based on importance and satisfaction scores or concept rankings. If you are running a qualitative study, you may ask a smaller number of customers to force rank or vote on ideas.)

PRO TIP **USE KILL METRICS AS PERMISSION TO FAIL**

Your test plan should specifically list what success and failure look like. Too many teams miss this part. If you're only looking for positive signals, you might fail to realize the outcomes are only incrementally better, flat, or even downright negative. Deciding what a failed outcome looks like before you start testing helps the team pivot when needed.

TABLE 5.2 HYPOTHESIS TEMPLATE

Theme	User Need to Solve	Hypothesis Description	Success and Kill Metrics	Results
Value Prop	As a user, I need the ability to accomplish [this aspirational goal].	We believe that doing X will result in Y.	We'll know we're right when we see [this signal]. We'll know we're wrong when we see [this signal].	*To be filled in after test*
Feature Set	#1: As a user, I need the ability to complete [this task or job-to-be-done].			
	#2: As a user, I need the ability to complete [this task or job-to-be-done].			
User Experience	As a user, I need the ability to do [this task or job-to-be-done] according to [these preferences].			

3. **Pick a testing method.**

   There's no singular way to test a new product concept. You can test concepts using low-fidelity, scrappy sketches (see Figure 5.7), different image treatments (see Figure 5.8) or stretch for high-fidelity, clickable prototypes. If you're not sure what methods and tools are available at your organization, check with your partners in research and design. Consider trying the methods listed at the beginning of this play or some of the research techniques listed in Chapter 4. If you're struggling with this part, here are a few tips to land a decision and prepare for testing.

   - Pick a method that will allow you to test competing hypotheses at once. Don't mix methods or test across different timeframes with different users—this will muddy your data. Also, consider whether you want to include or exclude your company's brand from the concept based on your learning objectives.
   - Ensure that you have a plan to collect, review, and analyze data before you launch the test. If you're using a qualitative method, consider ways to add some rigor. For example, you might consider gathering numerical data by asking participants to force rank or vote on concepts to understand relative performance of one concept over others.

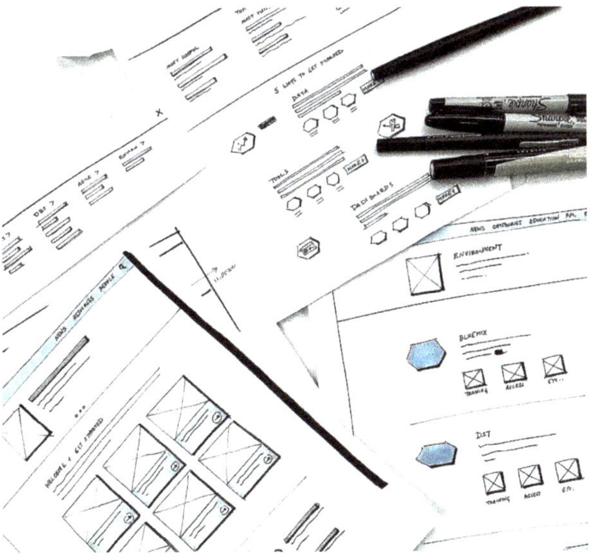

IMAGE COURTESY OF ALLY HWANG

FIGURE 5.7
Paper prototypes are a low-cost, low-effort way to get rapid feedback from users.

IMAGE COURTESY OF SURVEYMONKEY

FIGURE 5.8
SurveyMonkey provides quantitative methods to assess value proposition statements, logos, and images, as seen in these test ads.

4. **Create prototype(s).**

   Make sure to partner with researchers, designers, or engineers to build the prototype. If you're testing multiple hypotheses, allow your team time to design, iterate, and finalize distinct versions. Remember, it's good to test multiple hypotheses—just don't bundle them all in the same concept. For example, make a unique concept for each value proposition and present both to users. For rapid prototyping and testing, try out the tools listed here:

   - **Prototyping tools:** Cursor, Lovable (shown in Figure 5.9), Claude Code, Replit, Figma, Miro, Canva
   - **Survey tools:** Typeform, Google Forms, SurveyMonkey
   - **User research platforms:** UserTesting, User Interviews.

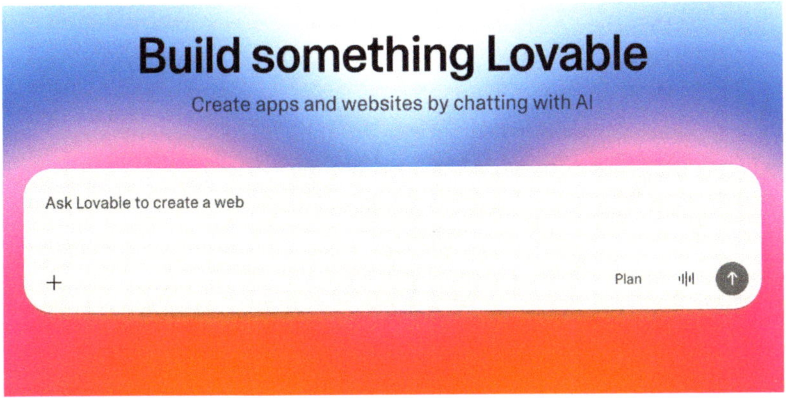

FIGURE 5.9
Generative AI tools, such as Lovable, can accelerate the prototyping process and supplement teams without design support.

5. **Run the experiment.**

   While some methods vary, generally speaking, you'll need to recruit users, deliver the test, and collect feedback. Reuse some of the plays from Chapter 4 to design, execute, and synthesize research results. That said, here's additional advice when testing concepts:

   - **Set expectations.** Let participants know they will see multiple concepts during the session.
   - **Reduce bias.** Randomize the order in which participants see concepts.
   - **Verify data collection.** Confirm that your collection tools work immediately after launch. This is especially important for digital tests or surveys.
   - **Maintain consistency.** Avoid mid-test adjustments. If you need to change concepts, reset the test and use post-reset data.

6. **Analyze results and determine next steps.**

   When you have enough data to draw conclusions, review the results. Start by filling out the results column in Table 5.2 and capture critical learnings. Then pull the team together and review these questions:

   - **Overall findings:** What concepts performed well and which ones didn't? What does the data tell us about the overall concept performance and individual hypotheses?

- **Target user findings:** Did we uncover new needs about our target user? Did we identify any new segments or categories of users based on these needs?
- **Value proposition insights:** What benefit statements resonated and which ones didn't? Does it vary by user?
- **Feature set insights:** What did we learn about each user success criteria? What functionality is table stakes, nice-to have, or not needed? Does it vary by user?
- **Experience insights:** What design treatments or design principles must be true to satisfy user preferences?
- **Iteration plan:** What could we add, change, or remove to improve the perceived desirability and utility of our proposed solution?
- **Next steps:** Do we have enough information to begin drafting solution requirements or should we run more tests to gather more data?

If the concepts didn't perform well or you want additional data before making a big investment in a new solution, plan for future tests and start this method over. Most teams do several rounds of concept testing and experimentation before they have enough feedback to draft solution requirements and nail down the initial feature set. That said, once your concept feedback is actionable and ready for the next step, continue with the next method.

LEARN MORE

CONSUMER PREFERENCES THROUGH SURVEYS

If you really want to geek out on data-driven concept testing, there are two additional survey methods you can explore, both based in market research practices:

- MaxDiff surveys, a survey method which asks users to rate best and worst options when presented with a given set of choices to assess trade-offs and preferences
- Choice-Based Conjoint analysis, a survey method which asks users to choose their preferred products and product attributes over a series of questions meant to mimic real-world trade-offs

PLAY 9

Building a Problem-Solution Fit Map

Why would a customer convert to our solution? What solutions solve the target customer's needs? What internal requirements does the company have before launching a solution?

After testing concepts, you'll have a strong sense of problem-solution fit—evidence that your proposed solutions address customer needs. (Product-market fit occurs later when you have evidence that customers will pay or adopt your solution.) This is a major milestone because it means you're ready to start translating the collective intelligence into product documentation.

Product documentation helps you distill product direction, scale knowledge about user needs, and maintain alignment as teams transition from discovery to delivery. As a PM, the most important documentation you own will be the product vision, product strategy, product roadmap, and requirements (covered in later chapters). But after wrapping up product discovery, you often need something lighter and action-oriented to help stakeholders understand product direction, including user needs and business needs.

Here's a method I use to create a simple, high-level product blueprint. It's not a substitute for formal product requirements, but it can help stakeholders quickly understand priorities and scope.

Use When

You need to distill research insights and communicate the high-level direction and scope for a new product or solution.

Average Time

2–8 hours, depending on the scope of the problem

Try This

1. **Assemble learnings about problem-solution fit.**

   Using prior artifacts and research readouts, compile and organize insights and decisions that answer the following questions:

   - What is the high-level goal you are solving for customers (the main job-to-be-done)?

- What is the new value proposition your company should offer to accomplish the goal?
- What are the critical user needs to address in priority order?
- What features will address each need?
- What preferences do users have about the overall experience in priority order?
- What are the key design principles that will accommodate these preferences?
- What is the rationale for each decision above?
- How do these decisions compare to competitors and existing solutions?

2. **Identify business requirements that must be true for products to launch.**

   Companies usually expect teams to meet specific criteria when launching a new product or system. For example, your company may require certain functionality to track analytics, stay compliant, or even recover from an outage. These qualities are nonfunctional requirements (NFRs) that help prevent or reduce risk when a solution is live. NFRs cover various categories as depicted in Table 5.3.

TABLE 5.3 COMMON NONFUNCTIONAL REQUIREMENT CATEGORIES

Category	Definition
Compliance	How well a system complies with regulations
Accessibility	How well users of all abilities can interact with the system
Performance	How quickly an application responds to requests
Security	How well a system protects against threats
Scalability	How ready a system is for scale or additional volume
Observability	How easily the business can observe user, application, or business metrics to measure performance or identify problems
Testability	How easily a business can test the application to ensure it functions as expected
Recoverability	How easily the system can recover from errors or an outage

NFRs are *business* requirements (not customer requirements), and PMs must include them in early product documentation to inform scope, estimates, and roadmap conversations. To identify nonfunctional requirements, ask your team, leader, or engineering partner if your company has existing standards, such as a "Definition of Done" or "Enterprise NFR" policy. Otherwise, pull your team together and identify essential criteria that must exist before you launch a solution.

3. **Organize everything into a Problem-Solution Fit Map.**

   A Problem-Solution Fit Map is a quick way to visually display what your solution should offer, including the value proposition, the key features, user experience guidelines, and business requirements. Consider this a companion document to the problem statements, personas, user research readouts, and concept testing results you've already captured.

   Many teams use a formal product requirements document to detail specific behavior about each solution, feature, and requirements (and I suggest you do the same). However, if you need a visual aid to quickly show the puzzle pieces of your solution and how they fit together, try filling out the Problem-Solution Fit Map as shown in Figure 5.10.

   When communicating a new feature set and scope, the most important part is showing stakeholders how the problem space (target user and user needs) connects to the solution space (value proposition, features, UX guidelines, and business requirements). Be specific and include evidence or rationale to help stakeholders quickly understand why and how you formed the overall solution recommendation. Use priority labels to show teams the difference between a Day 1 requirement versus functionality you can add later.

   You may need to adjust the Problem-Solution Fit Map to communicate your unique opportunity. If you're launching a brand new solution, you'll need to list many more features. Or, if you're serving multiple market segments, label which solutions and guidelines apply to different users.

Product Element	User Need to Solve	Proposed Solution	Priority	Rationale & Key Links
Value Proposition	*As a user, I need the ability to accomplish [this aspirational goal].*	Value prop statement #1	Must-have	Best performing value proposition statement in quant survey; appealed to 80% of target users
Feature Set	*#1: As a user, I need the ability to complete [this task or job-to-be-done].*	Feature #1	Must-have	#1 customer need; feature performed well in concept testing across 100 target users
	#2: As a user, I need the ability to complete [this task or job-to-be-done].	Feature #2	Must-have	Table-stakes feature across all competitors; users must have to switch
		Feature #3	Nice-to-have	Appeals to smaller, vocal subset of users within target segment
	#3 As a user, I need the ability to complete [this task or job-to-be-done].	Feature #4	Must-have	Under-served market need; top user satisfier in concept testing
User Experience	*As a user, I need the ability to do [this task or job-to-be-done] according to [these preferences].*	Design principle #1	Must-have	Minimizes cognitive load on customers & differentiates us from competitors
		Design principle #2	Must-have	Aligns to company brand; improves accessibility
		Design principle #3	Nice-to-have	Enables cross-platform interoperability
Business & Non-Functional Requirements	*#1 The solution must meet [this criteria] on Day 1*	Non-functional requirement #1	Must-have	Need this on Day 1 to meet compliance & regulation rules
	#2 The solution must meet [this criteria] on Day 1	Non-functional requirement #2	Must-have	Need this on Day 1 to track performance
	#3 The solution must meet [this criteria] before scaling	Non-functional requirement #3	Nice-to-have	Will need this before solution scales

FIGURE 5.10
Build a Problem-Solution Fit Map to visualize the puzzle pieces of your solution and socialize with stakeholders.

PRO TIP WRITING USER NEED STATEMENTS

There are different ways to represent user needs in product documentation. Regardless of the format, aim for consistency. Here are common formats you can test out:

- **Problem-centric:**
  - As a user, I need the ability to complete [this task or job-to-be-done], so that I can [accomplish this expected outcome].
  - As a [describe target user/stakeholder], I need a way to [describe task/job-to-be-done], so that [describe goal/desired outcome] occurs.
- **Solution-centric:**
  - A [user] needs a [feature/solution] in order to [accomplish/complete task].
  - Given [this circumstance occurs] when [this user] performs [this action] then the system should [respond in this way].

4. **Start sharing with partners and stakeholders.**

   After you draft the Problem-Solution Fit Map, it's time for a reasonability check with your team and stakeholders. Here's what to do:

   - **Pressure test the logic.** Share the Problem-Solution Fit Map with partners and stakeholders. Invite them to poke holes and ensure that the logic holds up.
   - **Identify the big rocks and dependencies.** Call out features that might require new capabilities or high levels of effort. Start listing potential dependencies on other teams, tools, or partners to avoid surprises later.
   - **Explore scoping options.** Your MVP goal should be learning, not scaling. Look for ways to keep scope tight.

   As you gather feedback from partners, find ways to track frequently asked questions and improve your overall documentation. These come in handy during feasibility assessments and major planning milestones.

Pulling It All Together

Product managers can't spend endless weeks overanalyzing problems because businesses actually want you to do something about them. Facilitating ideation, testing concepts, and assembling a Problem-Solution Fit Map helps you cross the chasm from defining problems to defining solutions using evidence-based techniques to validate or debunk assumptions. And user experience partners, guidelines, and systems accelerate your ability to deliver a useful solution.

The knowledge and methods in this chapter can certainly help you identify how to make a solution desirable and usable, but you haven't covered feasibility yet, which is where the rubber meets the road and constraints emerge. The next chapter covers tips to assess and fold in feasibility so you can (finally) crystallize the product strategy, roadmap, and vision.

EXPERT ADVICE WITH ALLY HWANG

WHAT DESIGNERS WANT FROM PRODUCT MANAGERS

To understand the relationship between designers and PMs, I turned to Ally Hwang, a design leader and expert with over 13 years of experience leading digital and physical design teams at Fortune 100 companies.

There's a lot of overlap in the role between product managers and designers. Both care about the user, the business, and creating solutions that drive value. Even user research tends to be a shared responsibility. That said, there are generalists and specialists. Don't make assumptions about the designer on your project. When you meet a new designer, ask them to share more about their design skill set, clarify roles on the team, and align on what deliverables they will create and own.

Once you start working with a designer, remember that a great partnership goes both ways and requires trust, inclusion, and context. Instead of going rogue and planning research in a vacuum or drafting concepts without a designer, include your designer early on when you're starting to frame the problem or design research. Designers can (and should) help shape the problem definition, especially as champions of users and desirability. There may even be times that design research suggests the team is solving the wrong problem or need. Trusting the designer and including them early during discovery can prevent the team from chasing the wrong problem altogether.

Strong product documentation is also critical for a great partnership. Throughout projects, designers need clear and consistent communication from PMs, including thoughtful requirements that document context on the problem space, the key players, and who to talk to for more information, user insights, and business goals. And when the team pivots, makes a trade-off, or converges on a new insight, great PMs update the documentation so that everyone has the latest context.

When it comes to defining solutions, sketching and simple prototypes are useful ways for PMs to communicate their ideas, but leave the higher-fidelity and interaction design decisions to the experts. When it's time for detailed design, it's better to be descriptive rather than prescriptive. Designers have inherent muscle memory and tons of expertise, especially when it comes to patterns, layouts, user flows, and selecting the right components. Trust them to make the best decision for users and validate through testing! Designers understand that research can be expensive and not everything requires testing. Lean on their expertise and make a team decision on what's worth testing versus what can ship right away.

CHAPTER 6

How: Defining the Approach

It's 10 a.m. on Tuesday. You're hosting a meeting to share results from last week's round of concept testing with customers. You've barely made it past your first slide when leaders start asking questions. "What's the level of effort to build this? When can you deliver it? Can you go *any* faster?"

You reach for your water and glance around the room hoping your tech partner will bail you out. Then suddenly it hits you—you haven't even talked to engineering yet.

Rookie mistake.

Once an idea looks promising, excitement spreads fast because leaders love tackling high-value problems. They also love knowing when they can expect results. But to answer *when*, you must first understand *how*.

To be clear, understanding how products are built isn't optional in product management—it's foundational. You can't craft a roadmap, manage value, or make intelligent trade-offs if you don't understand the technical aspects of your solution. And that's just scratching the surface. After you launch a product, technical fluency helps you plan upgrades, make judgment calls when things break, and drum roll please...actually *manage* the product.

Sometimes, technology *is* the product. For example, PMs who build APIs or platforms serve technical consumers and find ways for technology to unlock new revenue streams. But even PMs that lead consumer-facing products should understand technology strategy, data models, and capabilities that can drive differentiation in product, pricing, or distribution. Either way, technical chops are non-negotiable.

And yet some PMs insist they're "not technical" and focus on *what* products offer while ignoring *how* they work. Time for a truth bomb. Those PMs won't become strong product leaders. The strongest product leaders work across four dimensions: why, who, what, and how. Most PMs are comfortable with the first three, but the fourth one is where great PMs separate themselves.

The good news is that there's an entire workforce of technical experts who can help you learn. Once you know *what* to build, start leaning on your technical partners to figure out *how*. These are the questions that matter most:

- What will it take to make my product actually work?
- How much effort will it take to build it?
- How can I future-proof my product?

Technology 101

You don't need to write code to lead a product team, but you do need to speak the language. If you shy away from engineering discussions (or perhaps feel too embarrassed to ask questions aloud), find a trusted advisor or mentor. Every company has teachers, translators, and people who enjoy coaching—just ask around. Until then, consider this section a crash course in basic software engineering knowledge every PM needs.

Common Technology Partners Every PM Should Know

With over 25 million software engineers around the world, titles and skills vary across industries, companies, products, technologies, and teams. You'll likely engage several different engineering specialties as a PM. Here's a breakdown of the most common tech roles that PMs will encounter.

- **Tech leads:** Sometimes referred to as *engineering managers*, these individuals are usually senior-level engineers with people-leadership responsibilities. They determine the technical approach for a new solution, and they guide the engineering team when building, testing, and releasing the solution. Tech leads also work hand-in-hand with product managers and design leads. They can (and should) join user research sessions, help you scope and estimate new initiatives, and spot critical dependencies early. They also work closely with product managers and design leads by joining product discovery activities, scoping technical initiatives for the roadmap, coordinating estimates, and unblocking dependencies for the team.

  The relationship between a PM and tech lead is crucial. Both should attend all team ceremonies or meetings, including daily standups, planning events, requirements reviews, and retros. On high-performing teams, PMs and tech leads are co-equals. They align on the team's operating model, and they partner to ensure that the team delivers outcomes, not just outputs. When something goes wrong in production, the PM and tech lead are the ones accountable for communicating the status to leaders and determining a plan to resolve it quickly.

- **Software developers:** Sometimes referred to as *devs* or simply *engineers*, members of the development team are usually individual contributors who report to the tech lead or engineering manager. They bring the solution to life by coding, testing, and releasing it. Some engineers specialize in different domains, such as front-end development (what end users see), back-end development (configuring databases, building APIs, completing server-side work, etc.) or full-stack development (a combination of front-end and back-end skills). In addition, some engineers specialize in specific domains such as security, site reliability, machine-learning technologies, mobile applications, etc.

All devs should understand product requirements, help spot dependencies, and engage the PM to demo progress or ask questions. While the product manager works most closely with the tech lead, the PM should know and build rapport with every member of the dev team. In scenarios where the tech lead is not communicative or difficult to work with, having a strong relationship with dev team members is critical to keep work on track.

- **Software architects:** These individuals are experts in overall system design. They are responsible for understanding business and product goals and designing a high-level technical blueprint (including frameworks, data flows, and standards) for products and systems, so that teams deliver solutions that integrate easily with other systems, processes, or structures in the company. PMs should engage architecture experts before establishing a new solution or capability for advice on technology selection and system design. Architects can spot *one-way doors* (technical decisions that are much harder, or more expensive, to change later) and offer recommendations to help the company identify less risky approaches.
- **DevOps:** DevOps specialists are responsible for driving efficient, stable, and continuous software deployments. They accomplish this by blending technical expertise (infrastructure, automation tooling, and system administration skills) with operational expertise (communication, processes, and cultural norms) to ensure that teams can continuously integrate and release code without compromising system security or reliability. When PMs observe routine issues with software environment stability, code quality, testing tools, or software deployment, they should engage DevOps specialists to identify ways to improve the software development life cycle.
- **Delivery enablement partners:** Complex software delivery often requires additional coordination and expertise beyond engineering and product management roles, especially when multiple teams are involved. To fill this need, many companies

hire technical program managers, product operations experts, quality assurance testers, release train engineers, and other specialized roles that support software delivery and major tech-centric initiatives. It's common for PMs to interact directly with these experts; however, roles and responsibilities will vary across companies, so ask your leaders for clarity on how and when to collaborate with specialized functions.

That said, some companies have eliminated specialized delivery roles in recent years and shifted more responsibilities onto tech leads, product managers, and development teams. Without proper training, tooling, and staffing, many teams experience reductions in velocity, cross-team alignment, and product quality. If you're experiencing these issues, engage your product and tech leaders for support. Bring data to the conversation—including the number of defects, time spent on quality issues and dependency management, and velocity metrics—to build a case for additional staff and better tooling.

PRO TIP **FOUR SIMPLE WAYS TO BUILD TRUST WITH TECH PARTNERS**

- When starting a new product role, ask your tech partners to explain how the company organizes engineering functions and identify which teams they usually depend on to ship a new solution. (An org chart helps!)
- When inheriting an existing product, ask tech for a deep-dive on how the product works, including system architecture, underlying logic, critical services, differences across platforms, data models, service-level agreements, and monitoring tools.
- Involve tech partners early in the product development process, ideally during problem definition, so they can build stronger solutions.
- Never share a target release date without consulting tech partners first.

Technical Concepts Every PM Should Know

System qualities like speed, scalability, and reliability are more than technical specs—they directly influence your bottom line. When product quality is strong, companies can gain an advantage over competitors. But the story goes both ways. When products suffer from slow performance or frequent outages, teams start to see lost revenue and customer churn increase. Of course, technology is ever-changing and advancing, but every PM should master these fundamental concepts to build resilient, market-leading products.

Technology Strategy

A technology strategy is a set of decisions that outlines the technologies, capabilities, skills, tasks, and costs required to build, deliver, and maintain solutions. In most cases, engineering leaders own these decisions, but they form the strategy in close collaboration with business and product teams. As a PM, it's important to realize that companies often have multiple technology strategies in motion to address enterprise-level, product-level, or feature-level goals. For example, an enterprise technology strategy might standardize data tools or cloud providers for the company, but a product team may have some latitude in making specific technical decisions within enterprise constraints. Product teams should treat an enterprise technology strategy as "rules of the road" and look for ways to consume and reuse shared services, like identity management or payment services, instead of building their own version. Doing this will help you build solutions that cost less to maintain and can integrate seamlessly with other company systems.

Tech Debt

Tech debt represents the cost and quantity of future work that accrues when decision-makers prioritize speed over quality during the delivery process. For example, teams may opt to deliver a beta solution in-market with several known bugs because the perceived value of launching the overall solution is greater than the business risk or customer impact created by the bugs.

Product managers often make prioritization calls that involve reasonable, short-term technical compromises. Having *some* tech debt is normal; however, accruing massive amounts of unaddressed tech debt is *not* okay. When tech debt goes unaddressed for long periods of time, your team will spend more time in firefighting mode to *fix* or *replatform* the product versus actually *enhancing* the product. As a PM, never let tech debt amass to this point. If you build it, you own it.

Software Architecture

A software architecture plan represents a set of decisions that define the overall technical design and structure of a solution. Architecture plans also define the boundaries of subsystems and guidelines for how subsystems should behave when passing data from one system to another. Software architecture plans should communicate the relationships between systems or software components (see Figure 6.1).

Software architecture usually includes at least three layers of a solution: the presentation, business, and data layers. The presentation layer dictates what users actually see and is responsible for taking data from the business layer and displaying it to end users; this is where the user interface (UI) rules typically live. The business layer contains the logic and rules that dictate how the solution should behave given a specific context. Business layers often host most of the application's complex rules or the company's "crown jewels," so they are usually represented as multiple components or sublayers. The data layer fetches and provides raw data from databases required to support the business layer's logic and computations.

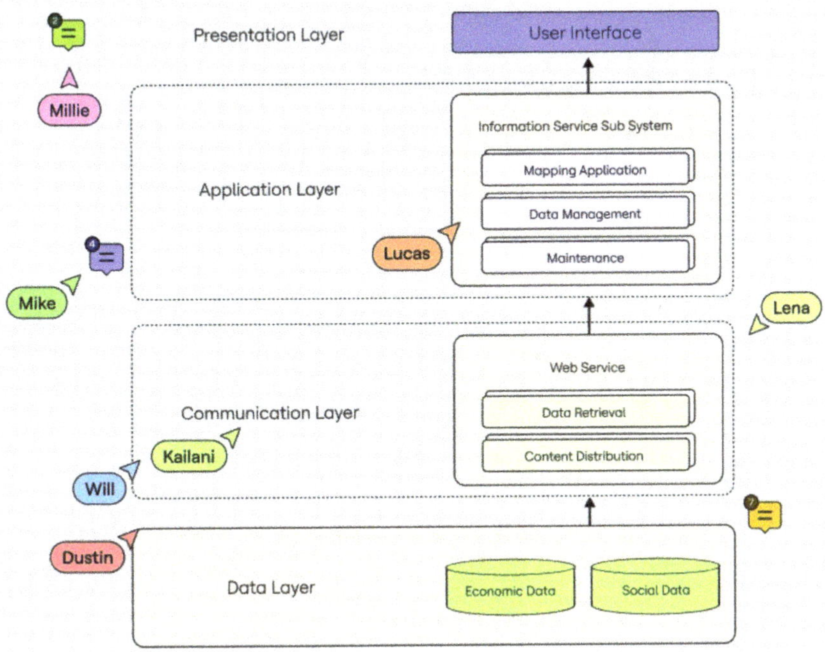

IMAGE COURTESY OF MIRO

FIGURE 6.1
Tools like Miro, Figma, or Draw.io help teams easily build architecture or sequence diagrams to communicate the relationship between systems in a solution.

PMs must ensure that their product's architecture supports critical functional and nonfunctional requirements, like maintaining data security, displaying a consistent user experience, enabling faster content updates, or even scaling quickly. When starting a new PM role on an existing application, you should ask the engineering team for a deep dive into solution architecture and get feedback on what's slowing them down or generating a suboptimized user experience because these problems will become future backlog candidates.

Microservices Architecture

Early software products typically used monolithic architecture, meaning all of the logic and business rules existed in a single, tightly-coupled unit. This approach made it difficult for teams to make small changes to the solution without updating, testing, and deploying a new version of the entire solution. Over time, companies began using microservices architecture, which emphasizes smaller units or systems that house their own business logic and process data faster, making it much easier to independently update, test, and deploy a change without needing to update other parts of the solution. Figure 6.2 shows how basic microservices can power a user journey including one microservice for user authentication, another for fetching and managing user preferences, and yet another for fetching and exposing product details. By managing these functions separately, teams can deploy changes faster with lower risk of inadvertently breaking other components.

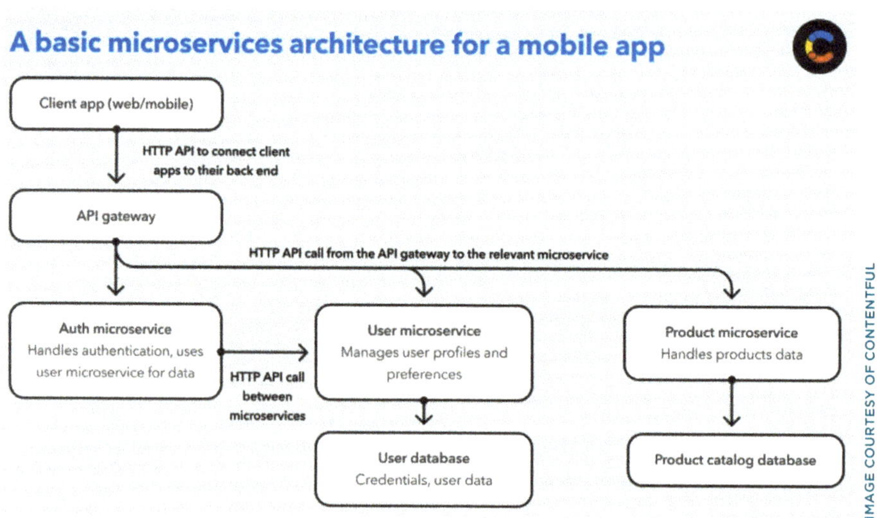

FIGURE 6.2
Both APIs and microservices are modern approaches to building efficient, scalable software—two tenets every PM should ensure to future-proof their products.

APIs

Application programming interfaces (APIs) dictate a set of instructions that software systems can use to communicate and pass data to each other, which speeds up product development and improves interoperability across systems. APIs can be public or private, and they make it easy to access functionality created by another team or expose your product's functionality to a partner. In some cases, APIs are an example where technology *is* the product because they create a new revenue stream. For example, instead of building an application that provides maps and directions, Google provides a public API that allows companies to integrate with their Google Maps service for a fee.

The Tech Stack

The tech stack is the collection of software systems and technologies that teams use to build, run, and maintain a website, mobile application, or other digital solutions. The stack metaphor allows teams to vertically arrange the technologies, frameworks, and programming languages used to create a solution, starting with technologies that support front-end development (like UX/UI component libraries) and ending with back-end development (like data storage and data processing) and operating systems (i.e., Windows, macOS, Linux). Figure 6.3 shows the common elements of a tech stack for a web application.

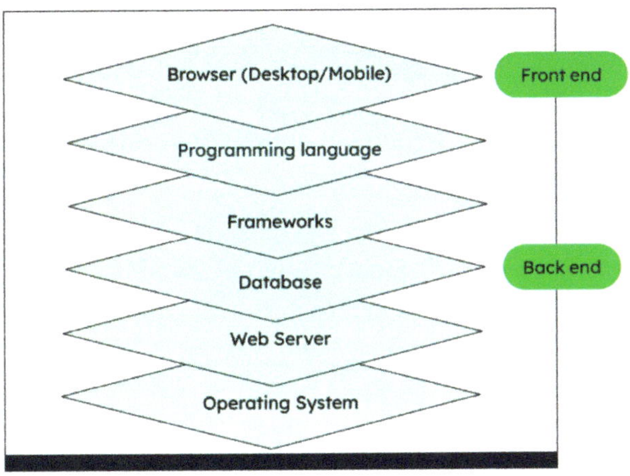

FIGURE 6.3
Tech stacks may differ across customer journey touchpoints. For example, marketing, sales, and customer service tech stacks often require different technologies.

Code Repositories, Branching, Merging, and Version Control

It's rare for PMs to code, but understanding how your team's code repository works, including how your team builds, tests, and merges code (and where they store technical documentation) is essential for understanding how code integrity and product quality is maintained. Code repositories or *repos* provide a centralized location where all the software code for your product lives, along with configuration files and documentation about your software project. In addition to housing the "source code" or master copy of the software, code repos provide branching functions that allow developers to clone the source code and work on a parallel copy. When developers are ready to merge or "commit" their changes into the master copy, code conflicts or broken functionality might occur. To resolve code conflicts quickly, code repos offer version control tools that show a commit history or log of what code changed, when it changed, and who changed it.

Infrastructure and Cloud Computing

Most PMs will work on platforms or migration projects at some point in their career. When you're tackling this type of work, it's important to understand "behind-the-scenes" tech and the costs of running it. In many cases, you'll need a solid understanding of your company's cloud computing approach and pricing structure to assess value.

Prior to cloud computing, most companies owned and operated all of their technical infrastructure "on prem," meaning on the premises of a physical location. This model required companies to purchase, own, and maintain servers and hardware solutions, while also paying individuals to maintain this infrastructure.

Today, cloud computing allows companies to outsource technical infrastructure assets by paying third-party providers (like Amazon, Microsoft, or Google) to deliver infrastructure and software storage solutions over the internet. This is possible because cloud providers maintain a global network of resources, data centers, and servers that store company data across multiple locations. This network and storage approach offers stronger data backup options, faster performance, easier scalability, and improved site reliability. This means when one server fails, or when companies experience a surge in digital traffic, product performance doesn't suffer.

Artificial Intelligence Explained

Artificial intelligence is not a single technology. It's a field of science that explores using a set of technologies to create solutions that can mimic human reasoning and cognitive ability, but at a much greater scale and speed as compared to humans. AI product management is the field of building or using AI-powered solutions to solve problems for customers. To build and leverage AI-driven products effectively, PMs need familiarity with a different set of technologies than traditional software requires. Some of these technologies include the following:

- **Machine learning (ML):** A subset of AI that offers a way for systems to learn from large amounts of data by using learning algorithms instead of traditional programming rules. To train new ML models, data science teams systematically provide training data to help learning algorithms uncover new patterns and produce better results. Common training methods include supervised learning (where the model learns by example from historical, labeled data to assist with activities like classification or diagnostics), unsupervised learning (where the model finds patterns among unlabeled data without guidance to support tasks like data clustering or detecting anomalies), self-supervised learning (where the system learns to create its own labels from unlabeled data for downstream tasks like text prediction), and reinforcement learning (where the system learns the most optimal strategy through trial and error). When training new models for AI-driven products, PMs should weigh factors like use case, costs, explainability, and model accuracy before selecting a learning method.
- **Large language models (LLMs):** Pretrained models that can process, understand, and generate human language to complete tasks like answering a question or summarizing a transcript. Some LLMs are text-driven, while others are multimodal, meaning they can process text, voice, audio content, or even code. ChatGPT by OpenAI, Claude by Anthropic, and Google Gemini are examples of foundational LLMs available to the public.
- **Retrieval-augmented generation (RAG):** A technique that allows LLMs to access, use, and cite authoritative knowledge from external data sources like research journals, newspapers, or statistics providers when generating answer outputs to users.

Because models are often trained on static data, RAG technology helps models generate up-to-date responses and reduce the risk of model hallucinations (when the model generates an output that sounds factual but is false).

- **Generative AI (GenAI):** A type of artificial intelligence that can learn from massive amounts of information to create new content, including images, videos, music, or conversations. Generative AI uses pattern-matching, probabilities, and statistics to generate outputs based on learned relationships from its training data. Common GenAI use cases include creating marketing materials, assisting with coding tasks, creating prototypes, or generating art (see Figure 6.4).

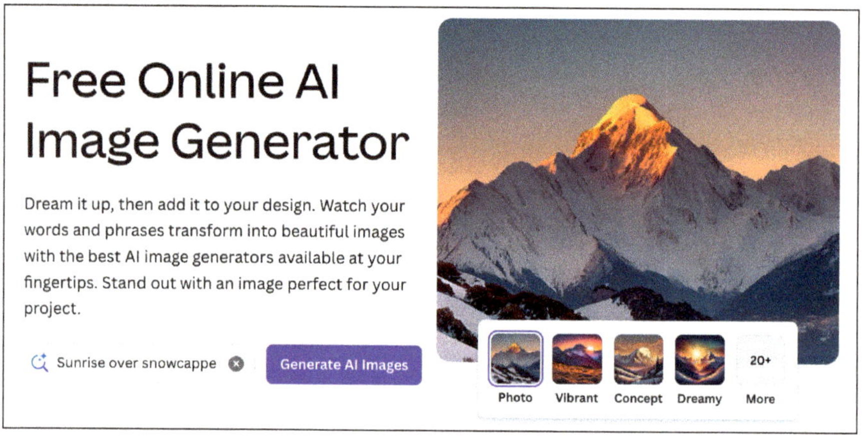

FIGURE 6.4

Canva's online AI image generator leverages generative AI to support content creation, a core use case for AI technologies.

- **AI agents:** Configurable, customizable software systems that leverage artificial intelligence technology to complete tasks on the user's *behalf*. These systems process video, text, voice, audio, and code to learn and conduct actions on behalf of the user's prompts. AI agents can power chatbot experiences, review and prefill your tax information during filing season, detect and prevent fraud on websites, interpret customer needs over the phone, or even book a dinner reservation for you based on your preferences.

The AI Product Manager's Starter Kit

As companies transform and invest in AI-driven technologies, the demand for AI product managers will rise. While many traditional product management skills still apply, PMs building AI-powered products must also cultivate specialized expertise. Here are six foundational skills you should develop now for a future in AI product management. (That said, this field is a moving target, so the top thought leaders are listed at the end of this section. Follow them to stay current on best practices.)

Understand AI Use Cases

AI isn't a silver bullet to product success. Whether you're leveraging AI-technologies or traditional tech, always start by understanding the problem space, including user needs, pain points, and emerging opportunities worth solving. Then you can assess whether AI's capabilities or other technologies are the right fit. You'll learn more on feasibility assessments in Play 10, "Assessing Technical Feasibility," but for now, it's important to understand a few areas where AI excels:

- Research, data analysis, and insight generation tasks, like summarizing text and data insights, predicting trends, or generating ideas during a brainstorming session
- Content generation tasks, like image, text, audio, or video creation
- Personalization and dynamic product tasks, like conversational assistants or product recommenders
- Workflow automation and productivity tasks, like code-assist tools or AI agents that act on behalf of users

To help unpack whether AI can support your specific use case or customer journey, OpenAI suggests starting with these three questions:[1] When do users perform repetitive, low-value tasks? When does a workflow slow down because users encounter skill gaps that slow them down? Where do users or teams navigate ambiguity during workflows, such as generating new ideas, insights or next steps? Once you answer each question, consider if and how AI could improve the user experience.

1 "Identifying and Scaling AI Use Cases" (white paper, OpenAI, 2025), https://cdn.openai.com/business-guides-and-resources/identifying-and-scaling-ai-use-cases.pdf

PRO TIP **THINK ABOUT LONG-TERM COSTS**

AI-driven products and experiences are not one-and-done deployments. These solutions require ongoing costs and operational support. If your company doesn't have the gusto to maintain and optimize AI-driven experiences for the long haul, or you aren't sure the benefit outweighs the total cost of ownership, consider alternative options.

Leverage Big Data

Data literacy is a core competency for all product managers (explored further in the next section), but it's especially vital for PMs building core AI technologies or AI-driven experiences. AI PMs should have a basic understanding of diverse data sources, data collection methods, and data processing steps. Literacy in these topics helps ensure that the product's data is clean and ready for model training and performance. In most cases, you'll partner with data experts to help with these decisions, and your primary role will be to manage value and risk along the way. For example, the PM should ensure that data sources and methods adhere to data privacy rules, meet user preferences, and mitigate ethical concerns while also balancing trade-offs across short-term costs, long-term costs, viability, and feasibility.

Optimize Models

Fine-tuning, RAG, and prompt engineering are three popular optimization techniques to improve a model's performance and the value it provides to users. Fine-tuning describes a resource-intensive process where teams retrain the model on new, specific datasets to improve accuracy for specific tasks (like enhancing a proprietary recommendation engine), whereas RAG optimizes a model's output by allowing it to retrieve and use external datasets without retraining it (great for tasks like featuring timely news snippets in search engine outputs). Prompt engineering, on the other hand, is a lightweight approach that allows teams to quickly steer pretrained models toward desired outputs by giving the model instructions instead of relying heavily on new datasets. Prompt engineering is one of the most hands-on ways a PM can improve model outputs, so let's take a deeper look at this approach.

Prompt Engineering

Prompts are instructions that a user provides to an AI model. Prompt engineering is the practice of using prompts to improve pretrained LLMs, so they accomplish a desired outcome. PMs that build or leverage

AI-based solutions can use prompt engineering skills throughout the product development process to guide the model and ensure that the model accurately and efficiently addresses specific use cases.

Prompt engineering techniques include zero-shot prompting (instructing the model to generate an output without providing an example), one-shot or few-shot prompting (instructing the model to generate an output and providing one or more examples to guide it), or chain-of-thought prompting (a technique used to help models break down complex reasoning requests into a series of intermediate steps). For the best outcomes, PMs should write clear, concise, and specific prompts to deliver the desired output when training models (see example in Figure 6.5).

```
# Identity

You are a helpful assistant that labels short product reviews as
Positive, Negative, or Neutral.

# Instructions

* Only output a single word in your response with no additional formatting
  or commentary.
* Your response should only be one of the words "Positive", "Negative", or
  "Neutral" depending on the sentiment of the product review you are given.

# Examples

<product_review id="example-1">
I absolutely love this headphones — sound quality is amazing!
</product_review>

<assistant_response id="example-1">
Positive
</assistant_response>

<product_review id="example-2">
Battery life is okay, but the ear pads feel cheap.
</product_review>

<assistant_response id="example-2">
Neutral
</assistant_response>

<product_review id="example-3">
Terrible customer service, I'll never buy from them again.
</product_review>

<assistant_response id="example-3">
Negative
</assistant_response>
```


FIGURE 6.5

Use few-shot prompting when steering a model to learn a new task.

Include the following elements[2] in your prompt to improve your training outcomes:

- **Instruction:** A specific task or instruction you want the model to perform
- **Context:** External information or additional context that can steer the model to better responses (the more specific, the better)
- **Input data:** The input or question that you are interested to find a response for
- **Output indicator:** The type or format of the output

Write Evals

Models are probabilistic, not deterministic, which means they can generate different outputs immediately and over time, even with the same inputs or prompts. This makes testing and validation less straightforward than traditional pass-fail test cases in software. While there are a few ways to validate models (including user feedback loops, automated testing, and comparative approaches), evaluations (or *evals*, for short) are another hands-on technique PMs should learn and use during model development, testing, and production monitoring.

Evals are tasks that can test an LLM's responses for accuracy, identify model weaknesses, or help PMs benchmark a model's performance against prior versions. PMs can conduct evals by writing validation logic, such as giving the model a prompt and checking the answer against expected outputs, or by asking the LLM to check itself for accuracy. When done consistently, the process of evaluation helps teams identify bias, inaccuracies, and risks in the LLM and determine when additional fine-tuning is needed.

Evaluate Third-Party AI Solutions

AI solutions are powering a new era of B2B Software as a Service (SaaS) products. Instead of investing time and resources to build their own solutions, many companies will purchase access to third-party LLMs and AI-based services to integrate with customer-facing products or use internally. Keep in mind, this industry is still rapidly shifting, so you should partner with technical experts to discuss

---

2 "Elements of a Prompt," Prompt Engineering Guide, retrieved 2025, www.promptingguide.ai/introduction/elements

how your team can leverage LLM-agnostic architecture patterns to manage cost, risk, and strategic optionality. And when you are ready to evaluate third-party solutions, be sure to conduct a robust assessment that spans functionality, viability, and ethics criteria. (Use Table 6.1 as a starting point and tweak it based on what matters most to your organization.)

TABLE 6.1 SAMPLE EVALUATION CRITERIA FOR THIRD-PARTY AI SOLUTIONS

Theme	Evaluation Criteria
Functionality	• What use cases, capabilities, and features do we need the solution to support right now and over time? • How well can this solution solve our stated problems? • How could we fine-tune or customize the solution to meet our needs? • What is the model's training history? • How well can the model's decisions be explained? • What are the solution's gaps, limitations, or downsides? • What is the level of effort to integrate with this solution?
Measuring Success	• What are success and failure criteria for the model or solution, including user experience, accuracy, quality, availability, volume, and latency goals? • What is our short-term plan to test the model or solution initially? • What is our long-term plan to measure return on investment?
Cost	• How are costs determined, and will this change over time? • What cost or budget constraints do we have? • What would renegotiation or exiting this partnership entail?
Data Usage and Model Ownership	• What standards must be true for model or data ownership, data privacy, and governance? • Does this vary if the solution uses third-party source data for training or if we customize it?
Regulatory, Compliance, and Ethics	• How can we ensure that this solution meets regulatory requirements, including security, privacy, and other applicable laws? • How well does the model meet industry-specific compliance guidelines? • How well does this solution align with our culture and stated ethics? • What guardrails exist to prevent harm to users?

Practice Responsible AI Standards

Anyone defining and building products is responsible for ethical decision-making. That said, ethical practices are increasingly critical to emphasize as companies create core AI technologies and AI-powered experiences. AI-driven systems use historical data to help organizations make important decisions at scale, like who is approved for a loan, whose résumé displays to recruiters, or even medical treatment recommendations. Without oversight, these systems can amplify systemic harm and biases when ethics are ignored. To prevent harm and safeguard user trust, product teams must *proactively* embed ethical practices across the entire product development life cycle.

You can practice responsible AI by establishing design principles rooted in issues that matter to users and stakeholders, such as bias prevention, accuracy, transparency, and privacy. Ethics leaders and industry professionals have established a new crop of frameworks and canvases (like the one shown in Figure 6.6) to guide these conversations. Many of them cover broader principles like societal impact, accountability, fairness, safety, inclusivity, and transparency. If you're starting from scratch or need to revisit ethical decision-making, use these tools to help your team assess the human impact of product design choices.

Talking about ethics is a great start, but product teams should define processes and decision frameworks that enforce responsible AI standards. At a minimum, ensure that your team embeds responsible practices across the following inflection points:

- **Model training:** Review diversity, reliability, and accuracy of datasets when choosing historical datasets to train models.
- **Data collection:** Review security, privacy, consent, and compliance practices when collecting and using new datasets to support the product.
- **User experience:** Consider fairness, transparency, and explainability of model outcomes and how to help users understand outputs. For example, design user experiences that explain how model outputs are probabilistic and can generate incorrect responses.
- **Error handling and feedback loops:** When training, validating, and optimizing AI-centric products, integrate human-in-the-loop practices to correct and improve model performance. For example, build feedback loops that allow users to report mistakes.

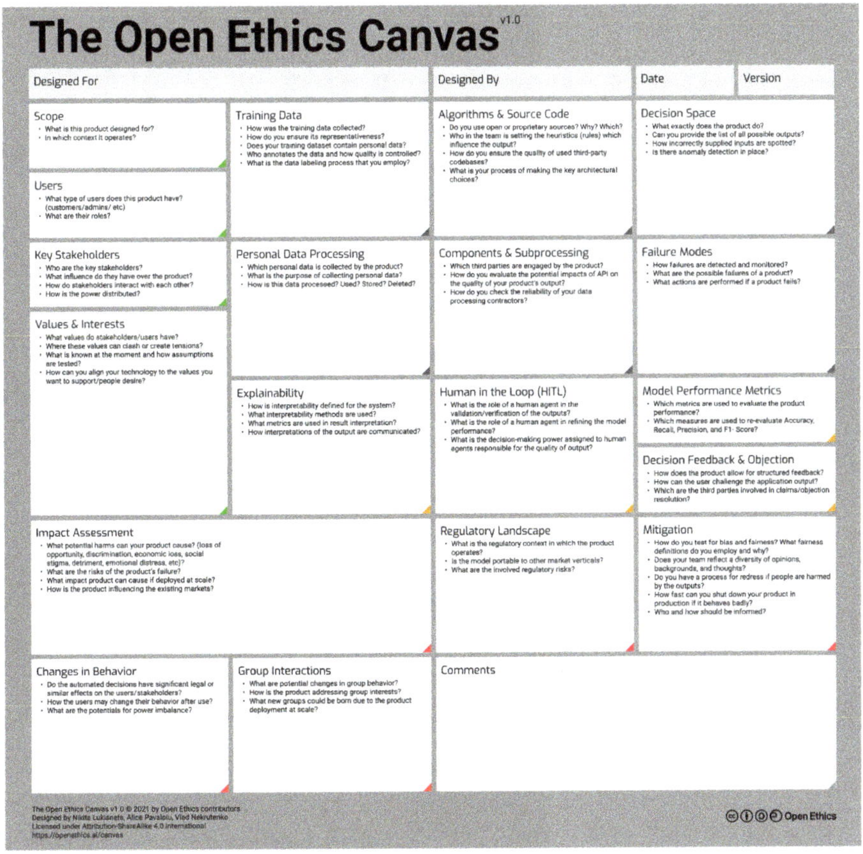

FIGURE 6.6

Open Ethics designed this canvas to help product teams build explainable, transparent technology products.

LEARN MORE

AI PRODUCT MANAGEMENT

For additional insights on AI skills, use cases, and concepts, check out these thought leaders and resources:

- Dr. Marily Nika, *Building AI-Powered Products: The Essential Guide to AI and GenAI Product Management*
- Andrew Ng, founder of DeepLearning.ai, co-founder of Coursera, and adjunct professor at Stanford
- Claire Vo, host of "How I AI" podcast, founder of ChatPRD, and chief product and technology officer of LaunchDarkly
- For information on ethics and responsible AI, check out the Open Ethics Initiative, the AI Ethics Canvas, and Microsoft's Responsible AI Principles

Data 101

Understanding data strategy, data architecture, and data partners are core competencies for every PM. Data is the lifeline of innovative products that customers love, and data teams are essential partners across many product development stages. Depending on the company, data functions may be referred to as *data science, data engineering*, or *product analytics teams*. Some companies embed data experts directly within product teams, while other companies centralize these functions under a common vertical or follow a hybrid model.

Regardless of org chart, PMs absolutely need a strong relationship with data teams. At a minimum, these teams can analyze data and answer specific questions to drive roadmap decisions, communicate results of an A/B test, or socialize overall product performance. In stronger partnerships, these teams help you think about data like a product itself, including what data to collect, how to collect it, where to access it, how to interpret it, and ultimately how to *derive value* from it.

Common Data Partners Every PM Should Know

It's a mistake to think that data partners are all the same. Just like software engineering, data teams usually include specialized roles that can help you answer critical business questions, build decision models, define your data strategy, and create the data products your team needs to generate insights at scale. Here are a few common roles you should know.

- **Data analyst:** Alternatively referred to as *business or product analysts*, data analysts typically analyze structured data sets and develop self-serve data products (like dashboards or visualizations) to answer specific business questions and inform roadmap decisions. On co-mingled product teams, data analysts may be embedded or aligned to a specific product area so they can develop deep knowledge of a product's data model and data ecosystem, including pipelines, reporting tools, and data constraints.

  PMs should partner with data analysts to define key metrics, build automated product scorecards that include business, customer, and platform performance metrics, and conduct data-backed storytelling that communicates product performance in the context of the broader company goals. PMs can also partner with data analysts to answer one-off operational questions like "What impact did the outage have on our product sales?"

- **Data scientists:** These specialists blend mathematics, advanced analytics, statistics, computer science, and industry knowledge to recommend enterprise- or product-level data strategies, answer strategic business questions and co-create products powered by AI and advanced analytical algorithms. They can perform detailed analysis to explain why something occurred (descriptive analytics), if something might occur (predictive analytics), and what the optimal outcome is for a company when specific situations occur. Similar to data analysts, they are adept at leveraging big data and advanced algorithms to power data-driven decisions.

  PMs should partner with data scientists to set a compelling data strategy for their product or feature, prioritize data quality, uncover patterns in existing data, design and analyze experiments, and build data-driven experiences, such as product recommendation engines, risk models, or dynamic pricing models using machine learning and AI-centric solutions.
- **Data engineers:** A subset of software engineering, data engineers are responsible for a company's overall data infrastructure. They design, build, and maintain data pipelines, databases, and systems that collect, process, store, and analyze data for reporting purposes, A/B testing, and AI/ML use cases. PMs partner with data engineers to ensure their product's data is efficiently flowing into the company's systems of record for tasks like financial reporting, updating the customer data platforms (CDP), or supporting bespoke reporting systems.
- **Machine learning engineers:** These specialists architect, launch, and optimize AI-powered solutions. They have many of the same skills detailed in each data role already listed, while also possessing AI-specific domain knowledge. As AI-based applications become commonplace in B2C and B2B products, PMs will engage more with ML engineers to identify AI use cases, leverage prompt engineering, co-create evals, measure solution ROI, and optimize products.

Common Data Concepts Every PM Should Understand

Data-driven—two words, four syllables, and way more intensive than junior PMs might assume. Being data-driven isn't just about using data to make decisions. It's about understanding how to make data useful and actionable. To do that, PMs must understand the data life cycle and ensure that their product has a data strategy that can fuel efficiency and product innovation.

Data Life Cycle

The data life cycle is a multistep process that starts with a data plan and ends with taking action, as shown in Figure 6.7. For each product, feature, or solution, PMs should start by making a data plan and identifying what data they need and why. Next, the PM should ensure that their product successfully generates, collects, and sends data efficiently through data pipelines where it's processed and stored for future use. Finally, the PM and data partner should analyze and interpret the data to answer critical performance questions like "Are we meeting our key metrics?" and then take action based on the answers. In large companies, this life cycle requires partnerships across multiple teams and several rounds of validation to ensure that the end-to-end data flow works as designed.

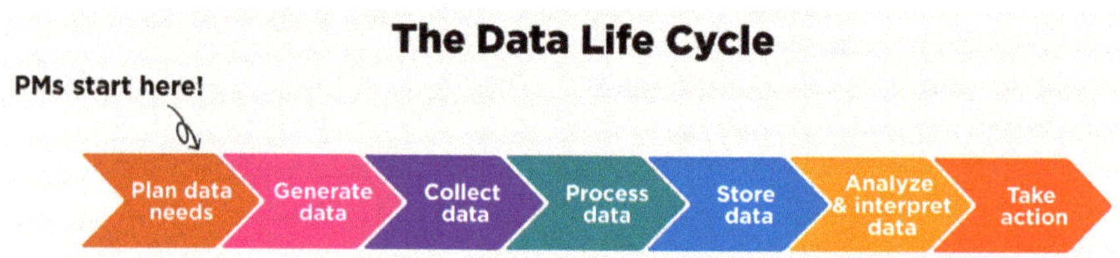

FIGURE 6.7
Data-driven PMs are intentional and collaborative across the entire data life cycle.

Data Strategy

A data strategy is a set of decisions that outlines how a team will collect, manage, govern, and use their data to accomplish business objectives and product goals. Data strategies can exist at the enterprise-level, product-level, feature-level or all three. In partnership with data experts, PMs can define and document data strategies that specify how the team will collect, analyze, and use data to fuel better roadmap choices, design better customer experiences, or help organizations maintain a competitive advantage with advanced analytics and data-driven solutions.

Data and Analytics Requirements

Customers, products, platforms, and businesses generate loads of data each time a user visits a digital application. By defining and implementing data and analytics requirements, product teams can observe exactly what happened over the timeframes they need, so they can report on metrics or triage issues. Whether you're building a brand-new solution or shipping a new feature, every PM should define what they want to track across the following categories:

- **Customer analytics:** What customers are doing (for example, what they are engaging with, when they visited, what device they were using, etc.)
- **Platform analytics:** How the platform or application is performing (for example, latency, uptime, error rate, etc.)
- **Business analytics:** How the business is performing (for example, revenue metrics, customer support tickets, efficiency metrics, etc.)

Data Model and Schema

A product's data model and data schema create the blueprint of what data is collected and how it's organized and defined in your product's database. Understanding and influencing a product's data model can help you design and build dynamic experiences. For example, if you're building a customer journey that helps customers complete their taxes, you might have a multistep flow where customers upload their tax documents and wait for an accountant to review them. To help customers see the status of their case, you'd need the ability to track if the customer uploaded their tax documents (or not) and if the accountant reviewed the files (or if they are still pending review). Tracking these statuses in your data model will make it easier to create a feature that shows a progress tracker to customers or possibly triggers messaging campaigns with updates about their journey.

PLAY 10

Assessing Technical Feasibility

What will it take to make my product actually work? How much effort will it take to build it? How can I future-proof my product?

With a firm grasp on data and technology concepts, you can appreciate why feasibility is a major factor in strategic product decisions. A technical feasibility assessment reveals whether your company can (or should) reasonably build, launch, and maintain a new solution with available resources. It also forecasts the cost and effort, which helps you build a complete product strategy and prepare for investment and head count conversations with leadership.

Technical feasibility assessments can be a formal multipage document or a one-page memo (check with leadership to see previous examples). Regardless of length, you should ensure that a feasibility assessment considers functional and nonfunctional requirements, provides estimates on the level of effort and costs, and documents assumptions. Here's how to get one started.

Use When

You need to evaluate the complexity, effort, and cost to launch a new product, major upgrade, or capability. (You won't need this level of detail for every new feature or small enhancement.)

Average Time

1–2 weeks, depending on partner availability and available documentation

Try This

1. **Gather and consolidate product documentation.**

   Before technical partners can tell you how to build something, you need to tell them why it matters, who your users are, and what requirements must be met. Compiling this should be easy if you've documented consistently across the product development process. Table 6.2 shows the artifacts you can use to streamline this for technology partners. (Hint: Make it easy on your partners: don't send them a bunch of disparate docs and expect them to read it all. Instead, consolidate the information into one, easy-to-understand document with links for reference.)

TABLE 6.2 INPUTS FOR A TECHNICAL FEASIBILITY ASSESSMENT

Feasibility Input	Artifact to Use
Business Opportunity	• Business Opportunity Workshop (Ch. 3) • Problem Statements (Ch. 3)
Solution Details: Value prop, feature set, and user requirements	• Problem-Solution Fit Map (Ch. 5)
Solution Details: Nonfunctional requirements	• Problem-Solution Fit Map (Ch. 5) • Enterprise NFR standards (if available)
Conceptual Design and User Experience Details	• Links to concepts, wireframes, mock-ups, prototypes, or specs • Design guidelines or principles
Success Metrics and Constraints	• Details about business goals or timelines

2. **Review the documents and feasibility assessment outline with technical partners.**

   Engineering teams are generally very busy coding, testing, releasing, and managing existing solutions, so don't expect to send your product docs via email and get a polished assessment back in a few days. Feasibility assessments usually require several conversations across teams to establish context, clarify gaps, and lock in details. After you've compiled product documentation in Step 1, do this:

   - **Schedule a meeting with your technical partners and send a pre-read in advance.** Include partners in software engineering, data science, analytics, data engineering, and architecture functions. (And don't forget to include your design lead.) You'll need at least 60 minutes.
   - **Host a Q&A discussion with your tech partners.** Ensure that your technical partners have all the inputs needed to assess feasibility. At the beginning of the meeting, ask everyone to read the document quietly and keep track of their questions. Once everyone reviews the doc, it's time for Q&A. Take notes and record the call, so you can send it out later.
   - **Review and agree on the technical feasibility assessment outline.** Review and agree on a rough outline for the feasibility assessment, using the categories in Table 6.3 as a guide. Get specific units of measurement. For example, do you want to

see the estimate in total development hours, capital expenses, or both? Also, determine whether additional artifacts are required, such as an architecture diagram.

- **Agree on a due date, owner, and schedule for final review.** Once you've agreed on the next steps, send out the agreements and keep track of the progress. Engineering teams are usually busy. Put a goalpost on the calendar and check in to ensure that the team is on track.

TABLE 6.3 OUTPUTS OF A TECHNICAL FEASIBILITY ASSESSMENT

Feasibility Category	Questions to Answer
Implementation Plan: How can we build this?	• What new and existing technologies will it take to build this? • What services must we integrate with? • What is our proposed architecture plan? • How might we sequence the work to build this?
People: Who will build it?	• What talent or skills do we need to build this? • What teams and roles are we dependent on?
Effort and Costs: How much effort will it take to build it?	• What is the total time to build it? • What is the breakdown of time per feature? • What is the total cost to build it? • If we have X resources, we can accomplish this by Y date assuming the following is true. Our confidence level is Z%.
Future Considerations: What else should we consider?	• What will it take to support and maintain this over time? • Would we be able to scale it? • If we need to pivot, what could we reuse? • What is the biggest risk/unknown factor?
Assessment and Alternatives: What is the official assessment on cost, effort, and alternatives?	• To accomplish X over Y timeframe, we will need Z resources. The cost will be ABC. • Should we build, buy, or partner? • What trade-offs could we make to change our budget or timeframes?

3. **Review draft assessment together, finalize, and begin planning.**

   When ready, host a walk-through of results. During the walk-through, the product team should:

   - **Think like an executive.** Review the draft through the lens of a CPO or CTO. Anticipate and ask tough questions to find gaps and improve the draft before sharing it with leadership.

- **Fix gaps and finalize the assessment.** Incorporate the assessment into your strategic product documentation. Keep the final document in a location the team can access.
- **Start preparing for next steps.** Use the assessment to clear blockers or information gaps. For new initiatives, use it as part of funding pitches. For existing products, use findings to refresh your strategy and roadmap.

HOW TO TRANSITION INTO DATA PRODUCT MANAGEMENT

If you find yourself falling in love with data products, check out data product management roles—a specialized function where PMs own data assets, data pipelines, data schemas, self-service dashboards, data solutions, and forecasting tools. For PMs considering a transition into data product management, consider this advice from data product expert, Derek Schwartz:

- **First, establish a robust data product mindset.** This includes learning more about data schemas, data models, how to optimize pipelines for speed and granularity, and how to create durable reporting tools that don't break!
- **Then learn the technical and architectural side of data processing.** This side includes learning how to manage end-to-end data processing and infrastructure costs to drive ROI for an organization. Dig into pros, cons, and costs of various data processing tools and technologies offered by larger hyperscalers like Google, Amazon, and Microsoft, as well as data platform providers like Databricks, Snowflake, and Palantir. These can cost hundreds of thousands of dollars, so think of this like your P&L.
- **Start familiarizing yourself with AI/ML methodologies that can fuel data-driven decisions.** This includes studying up on Bayesian and probabilistic approaches for A/B testing, as well as studying propensity models or next best action modeling. You should also understand how and when LLMs can transform business questions into executable structured query language (SQL) to answer business questions that fixed dashboards cannot.
- **And finally, learn to master data-backed, executive storytelling.** Sometimes less is more when giving data to executives. Focus on the essentials first.

EXPERT ADVICE WITH DEREK SCHWARTZ

WHAT DATA PARTNERS WANT FROM PRODUCT MANAGERS

Derek Schwartz is an experimentation expert, the head of data science and analytics for Intuit's Identity team, and one of the most brilliant data product thinkers I know. Here are Derek's three rules to build data fluency as a PM.

Rule 1: Treat your product's data as an asset from the start.

Before launching a product or new experience, define all the product and business questions. Then document data and analytics requirements from the onset with your engineering partners. If the team feels stuck, start by asking these questions and leaning on data experts for support:

- What do I want to track, monitor, and analyze after my product launches?
- What are my data quality requirements?
- How will my software architecture support my requirements?
- What are the data self-serve and analytics use cases required from my product?
- How will people access this data, and who am I dependent on for that?

Rule 2: Insist that your product has a clear data strategy and data model.

A data strategy defines what you want to collect and why. A data model defines exactly how the data is structured, organized, and stored in each system. Having a clear understanding of your data model gives you flexibility in the experience that you want to create. PMs don't need to be experts at coming up with the data model, but they should identify that a data model is needed and engage data partners to think about the various use cases that inform the data model's design. (One word of caution for PMs: don't rely too much on clickstream data, given the data loss due to incognito browsing and regulations. For the metrics that are critical, consider instrumenting server-side events so you have 100% confidence in the numbers.)

Rule 3: Be specific when engaging data partners.

Just like you, data partners have deep backlogs. Engage with them strategically versus tactically so they can help you drive the outcomes you're seeking. For example, when you're asking them for support, such as a new data pull, a dashboard, or custom analysis, they need as much context as possible to prioritize their work. Describing the business opportunity, the *types* of questions you have, and how your request will drive business decisions helps them help you.

Pulling It All Together

Product managers who focus on *what* the solution offers without understanding *how* the solution works cannot become great product leaders or build strong judgment. That's because technology strategy *is* product strategy. Data strategy *is* product strategy. In fact, any decision about your product that articulates how it will achieve its objectives is product strategy.

Exceptional PMs understand that technology choices and data models are not just methods to distribute a product, but rather they are core assets in your product's value chain. They build strong relationships with partner teams, and they are nosey enough to develop an informed point-of-view on these topics. Making smart choices from the onset will enable speed, agility, and data-driven decision-making for your team after launch. Making uninformed choices can expose you to future risk, like slow delivery schedules, security vulnerabilities, or quality concerns—all of which lead to incremental damage and mounting tech debt.

With expertise from tech and data partners and a technical feasibility assessment in tow, PMs can confidently move forward and begin assembling the fundamental artifacts that crystallize product decisions and keep stakeholders aligned—the product vision, the product strategy, and the roadmap.

EXPERT ADVICE WITH LEON FEUERSTEIN

WHAT ENGINEERS WANT FROM PRODUCT MANAGERS

Leon Feuerstein, senior director of software engineering at a top U.S. bank has over 25 years of experience collaborating with product managers and tech teams. In his experience, PMs that focus on creating high-trust teams drive the biggest impact for businesses and customers.

The most effective product managers create trust with engineers, ensure that the team's goals are in complete lockstep, and avoid unintended surprises. And they do it by practicing three key habits:

#1 Study Technology: Once you understand customers' needs and product goals, study the technology that will bring your solution to life at whatever depth it takes to be conversational and build trust with tech partners. Engineers want you to understand their challenges, so it's important for you to understand the tech stack, third-party services, infrastructure tooling, and cloud services they interact with every day. These vary by company so revisit this exercise every time you start a new gig. Engineers also want PMs to have a basic understanding of deployment strategies and what it takes to release the solution. That means you should learn about development pipelines, environments, release processes, testing methods, etc. When PMs understand these concepts, they're able to unblock the team, understand dependencies, and communicate updates more effectively to nontechnical stakeholders.

#2 Practice Inclusion: Engineers want PMs to be inclusive. Most PMs know they should communicate business goals to tech partners, but the best PMs find ways to include engineers early in the product development process. Engineers who observe user research produce better solutions because they're exposed to the types of problems they need to solve versus solving a very specific problem from a single user story. Early exposure to customer research helps engineers design systems that have a stronger runway for scaling or pivoting later.

#3 Establish Co-mingled Backlogs: Devs want PMs to create one realistic, single backlog that displays product priorities alongside technical needs for the sprint team. PMs should think of owning products like owning a car. You don't own a car and expect zero maintenance—you take care of it, so things don't break down. The same applies to products and systems, and the team needs your support to ensure that the system doesn't fall behind just because these tasks aren't visible or super exciting. On the other hand, if you do make a decision that results in tech debt, then own it. Engineers understand trade-offs happen but create space in the backlog to proactively address tech debt. This helps the team avoid excessive firefighting later.

CHAPTER 7

Establishing a Plan and Getting Buy-In

If you're a solo PM at a startup or the only product person at a large company, you already know who owns product artifacts and alignment duties. (Look in the mirror.)

But the lines get blurry when you're a PM within a larger product team, and you might wonder who really owns the product vision, strategy, and roadmap, or how they're made. That's because PMs on larger teams often own a *slice* of a product portfolio—such as a critical platform, feature, or service—rather than an entire product. In these organizations, planning processes may be unclear or invite-only, if they exist at all.

But one truth holds for every PM: no scope is too small to build a vision, strategy, success metrics, and roadmap. Owning a critical system, feature, or service doesn't make you less of a PM than someone who owns an end-to-end product. Sure, the breadth varies across the two, but every PM is still accountable for thinking strategically, aligning stakeholders, and creating value.

So, if you haven't had a chance to define a vision, strategy, metrics plan, or roadmap yet, here is some advice: stop waiting for someone to ask you. You don't need permission. In fact, showing initiative with a smaller scope proves that you're potentially ready to tackle larger, more complex portfolios.

Consider two scenarios. Product manager #1 leads an innovation team that wants to pitch a new product concept but needs funding and executive sponsorship. Product manager #2 runs a small sprint team focused on the customer support experience and wants to recommend a strategic pivot in capabilities. Both need to follow the same steps and answer the questions that matter most:

- How will this solution benefit our business and how will we measure success?
- What's the product vision?
- What's our strategy to succeed in-market?
- What will we launch, when will we launch it, and what impact will it have?
- Do I have support to pursue this?

PLAY 11

Projecting Impact and Defining Success Metrics

How would this solution benefit our business? How will we measure success?

After weeks of strategy workshops, user research, concept testing, and gathering technical feasibility inputs, you'll eventually hit a pivotal moment and ask yourself "Should we actually build this?"

In some cases, the answer is obvious. A high-impact, low-cost feature is probably worth building if customer demand is strong. But bigger bets, like launching a new product, pursuing a redesign, or migrating customers to a new solution, require analytical rigor.

When time, effort, and costs are high, you'll need a reliable way to project impact and measure success. In well-resourced companies, you'll partner with finance, marketing, and data for this step. Some PMs, however, don't have this level of support to make everyday priority calls for their roadmap and backlogs. Instead, they need quick, straightforward ways to estimate value and track progress. If you're flying solo or working with partner teams, this method can help you project expected impact and build a measurement plan that feels right.

Use When

You need to assess the financial benefit from a new solution, select key metrics, and define a plan to track results.

Average Time

5–8 hours

Try This

1. **Estimate the value of your product, feature, or solution.**

   It's rare for individual PMs to manage their own profit and loss (P&L) statements for a specific product or business. In most organizations, P&L ownership sits with executives or heads of product. That said, most of your executive sponsors will have P&L ownership and expect individual PMs to consider benefits and costs when making priority calls and requesting additional support. To do this, PMs can use a simple return on investment (ROI) formula to estimate how much value you'll get from a solution compared to the cost of building and operating it. ROI is typically expressed as a percentage and looks like this:

Estimated Return on Investment (ROI) = (Estimated Net Benefit / Estimated Cost) x 100

To estimate ROI for your product or feature, you need to identify the following information:

- **The estimated cost to build a solution.** Add up all known costs to design, build, and operate your product over a given timeframe. Use cost inputs from your technical feasibility assessment. Also, gather cost inputs from design, product, marketing, and other teams if necessary.
- **The estimated financial benefit your solution will create.** Estimate revenue gains or cost savings that your solution will generate. For example, if your solution is a new product or feature that will boost conversion or retention, translate the expected increases into financial impacts such as new revenue dollars. If you have a solution that is focused on efficiency, translate the efficiency gains into expense savings, salary savings, tool savings, etc.
- **The estimated net benefit.** Start with the estimated financial benefit and subtract the estimated cost to build your solution. The remainder is the net benefit.

PMs can use the simple ROI calculation when assessing the value of a new product or comparing several ideas at once, as shown in Table 7.1. Using this approach can help you identify known costs, project financial benefits, and estimate overall value for conversations with stakeholders and your team. Of course, the ROI output relies heavily on facts and data-driven assumptions or insights you've collected during product discovery activities. In situations where you need detailed financial models, consider coordinating pre-work with partner teams in finance, marketing, tech, design, and data. Examples of pre-work could include completing a business model canvas together, exploring pricing tests, or even completing pro forma models.

TABLE 7.1 EXAMPLE SCENARIOS AND ROI CALCULATIONS

Scenario	Example
New Product Launches	• You're evaluating whether your company should build a new subscription-based product that provides investment advice to hobbyist investors, one of your target markets. • Your new subscription product will cost $250,000 to design, build, staff, and operate over the next two years. • You expect to generate $2,000,000 in revenue over the same timeframe because you expect to acquire 100,000 customers that each pay $20 for an annual subscription. • The estimated ROI for your product would be as follows: ($2,000,000-$250,000) / $250,000) x 100 = 700%.
New Feature Launches	• You're considering two new feature enhancements for an app that supports residential security devices. One feature would improve sales, and the other would improve renewal revenue. • Feature 1: Costs $5,000 to build and will increase sales revenue by $45,000 over the next year through a referral program. • Feature 2: Costs $30,000 to build and will increase renewal revenue by $150,000 over the next year. • Using the ROI formula, the estimated ROI for Feature 1 is 800%, and the ROI for Feature 2 is 400%.

2. **Assess overall viability of the solution.**

   While ROI is a helpful tool, don't use it as the sole factor when deciding "Should we build this?" Instead, use it as a critical data point in a broader conversation about the solution's overall viability. You can pressure-test overall viability by asking these questions:

   - How confident are we in our underlying assumptions about costs and financial benefits?
   - How well does this solution align with our company's stated goals for growth, profitability, or other objectives?
   - How does this solution's expected ROI compare to other solutions in our pipeline?
   - What is the payback period and when might we expect to break even on this investment?
   - If we pursue this solution, what else do we have to trade off?
   - If we don't pursue this solution, what impact might we see?
   - What is our overall recommendation: pursue or abandon this solution?

In low-risk scenarios, you could make this decision on your own or with your design and tech lead. When the solution requires additional resources and investment dollars, gather a set of key decision-makers, including your product leadership and executive sponsors. In either case, make sure to document the assumptions and decisions to keep the team aligned.

3. **Determine metrics that matter.**

   If you plan to move forward with a solution, select key metrics to track. Pick metrics that best represent the *primary* business goals and objectives of your solution and can show actual progress toward the goal. These metrics will become key performance indicators (KPIs) that you'll report consistently. Along with KPIs, PMs often track *drivers* of KPIs such as platform health, product efficiency metrics, and customer success data. Table 7.2 illustrates common metrics product teams track as a solution matures across the product life cycle.

TABLE 7.2 KEY METRICS ACROSS PRODUCT LIFE CYCLE

Business Goal	Metrics to Track
Validate Product-Market Fit with MVP	• Customer inquiries • Conversion rate • Visit, usage, and duration metrics • Customer churn rate • Customer retention rate • Customer satisfaction scores
Increase Product Adoption and Customer Growth	• Everything above, plus: • Lost shopper rates • Activation rates (onboarding, enrollments, etc.) • User task success rates • User task completion times • Market penetration % • Customer acquisition cost • Average revenue per user • Monthly or annual recurring revenue • Customer lifetime value
Scale Operational Excellence and Protect the Competitive Moat	• Everything above, plus: • Market saturation % • Customer tenure rates • Customer support and escalation rates • Expansion revenue (new revenue from existing customers or new capabilities/services) • Platform performance metrics (defect rate, application speed, throughput, uptime, etc.) • Time spent on bugs vs. enhancements • Total cost of ownership

4. **Set targets for key metrics.**

   Target setting is a bit of art and science. Targets should be aspirational, but realistic and achievable for the team. When setting targets, you should:

   - Use industry benchmarks, historical data, or current solutions to establish baselines.
   - Be realistic about the level of support you can dedicate to improve metrics.
   - Establish targets over specific timeframes (such as quarterly or annually).
   - Involve your team and stakeholders for feedback and alignment.
   - Adjust the target when new evidence suggests you should.

5. **Identify a plan to track performance (and update requirements if needed).**

   After clarifying ROI, success metrics, and targets, revisit your data requirements with technical partners and formalize a plan to collect and report on metrics. Do this by:

   - Reviewing the metrics with your technology and data partners
   - Identifying if your data model and current data tools can properly track the metrics
   - Updating nonfunctional requirements if you need a new solution or custom build to track a specific metric or build a new reporting tool
   - Documenting the source of truth (report, tool, graphic, etc.) you will use to report on metrics once your solution launches (see Figure 7.1)
   - Identifying data owners and accountable POCs (proofs of concept) for pre- and post-launch reporting needs

Primary KPIs	Average revenue per customer	Customer retention rate	Page Load Times
Target	$225 over 12 month trailing period	15% by Q3	p99 at < 2 seconds or less
Secondary KPIs	Conversion rate Average order value	Repeat purchase rate Product return rate	Visits per hour
Source of truth	Digital Sales Report - Tableau	Customer Churn Report - Power BI	App Latency Report - App Insights

FIGURE 7.1

Build a product scorecard like this to align stakeholders on key metrics, secondary metrics, and sources of truth before you launch a solution. Once the product launches, you can use this for routine performance updates with stakeholders.

PLAY 12

Defining a Product Vision

What is our product vision?

So far, you've seen methods that help PMs define and articulate why, who, what and how in the product development process. These techniques produce the essential building blocks that transform an ambiguous product discovery journey into a crisp solution concept that's vetted for desirability, feasibility, and viability.

But great product concepts never see the light of day without support from others. And by now, you're probably keeping a mental list of risks, major dependencies, and serious concerns (including costs, staff, or difficult stakeholders) that could make it hard for you to launch the solution unless you can build excitement, attention, and alignment with people and teams who can help.

When you arrive at this point, it's time for a major mindset and behavior shift. Instead of hosting deep dives on the research, concept testing, and problem-solution fit discussions, it's time to start a full-court press focused on gaining support, aligning stakeholders, and influencing key decision-makers so you have the resources required to fuel product delivery. And it starts with having a strong product vision.

Product visions compel leaders, stakeholders, peers, and teammates to care about a solution. They describe the future that your product will create for consumers and help stakeholders understand the primary benefits your product will offer and how it will support company goals. Great visions inform strategy and convince people to fund your product, join your team, and contribute resources to your cause. They also help new teammates onboard to your initiative and quickly understand strategic direction. And thankfully, they are not hard to craft.

Use When

You need an inspiring way to align stakeholders on the product's purpose and future-state.

Average Time

1 week or less, depending on stakeholder availability

Try This

1. **Gather context about the solution.**

   Writing a vision is easy if you've driven the entire product discovery process from scratch and you've documented insights along the way. However, most PMs aren't shipping new products every day. In fact, most PMs support *existing* products and some of those products need a first-time vision or a vision refresh. If that's you, the first step is to observe and gather context about the initiative or existing product. Kick off a listening tour and ask leaders, peers, and stakeholders the following questions:

   - Who does this product serve?
   - Why does/should this product exist?
   - What key pain points does/should this product solve?
   - What are the key benefits this solution should offer compared to alternatives?
   - Why should customers choose our solution now and in the future?
   - What have we learned since launching our product that should influence future direction?
   - What change should we create for customers and our company in the next 1–3 years?

2. **Take a stab at drafting a vision statement.**

   A product vision statement should be inspiring, opinionated, and concise. You can tackle this step with others (recommended) or by yourself, but the key is to be bold and pithy. Give everyone a template that resembles Figure 7.2 and start filling in the blanks. This step might take a few attempts until you land on wording that feels right.

PRODUCT VISION MADLIB

Our ____________ (product name) serves ____________ (target customers) who need to ____________ (accomplish this goal) by providing a ____________ (product description) that offers ____________ (key benefits).

Customers choose us over ____________ (alternatives) because of ____________ (unique selling points). Our product will allow us to capture ____________ (valuable outcome) for our company.

FIGURE 7.2

Try multiple versions of this template until you land on a compelling blurb that's ready for feedback.

3. **Share with trusted partners and integrate feedback.**

   Once you've drafted a vision, it's time to shop your vision statement around with a set of trusted stakeholders. If you already partnered with close peers in Step 2, try getting feedback from those outside of your normal circle. Colleagues across various teams, including creative professions such as sales, customer support, brand, marketing, and content strategy, usually have a great handle on customer needs, creative language, and key selling points. Ask them if your vision resonates and if it's clear. And of course, use their feedback to iterate and improve.

4. **Start the roadshow and experiment with formats.**

   Every product artifact serves the same purpose: to influence others to adopt a shared point of view. After you've got the product vision, you can start sharing this with your boss, leaders, and stakeholders in checkpoint meetings (or you may choose to wait until you have a strategy and roadmap to share). You can also share this with other teams you'll depend on for resources and support during product delivery.

   As you broadcast the vision, keep this in mind: you aren't truly influencing anyone until you start hearing other people use and refer to the same talking points and artifacts that you use. To keep things fun and fresh, try experimenting with different formats, such as a video featuring the team or customers, a prototype with a voice-over, or another memorable medium. Unless you're pivoting or changing strategy, product visions should be relatively stable over time, but your format doesn't have to be.

PRO TIP CRAFT A BOLD PRODUCT VISION

If you're feeling stuck, try looking at a few examples for inspiration:

- **Intercom:** Fin is the only complete, fully configurable AI agent system in customer service—empowering support teams to customize, test, and continuously improve Fin through a no-code user experience that anyone can manage. **https://fin.ai/**
- **Slack:** Slack helps you make work life simpler, more pleasant, and more productive by providing an AI-powered platform for work that brings all of your conversations, apps, and customers together in one place. Around the world, Slack is helping businesses of all sizes grow and send productivity through the roof. **https://slack.com/about**
- **Miro:** Miro provides a visual workspace for innovation that enables distributed teams of any size to dream, design, and build the future together. Simply put, they are here to help you create the next big thing by providing a powerful space where your team can collaborate seamlessly—even if you aren't in the same room. **https://miro.com/what-is-miro/**

PLAY 13

Defining a Product Strategy

What's our strategy to succeed in-market?

When shopping around a product vision, it's helpful to include a strategic plan describing how you'll accomplish it. This is where product strategy artifacts shine. Think of them as a companion to the product vision—and yes, crafting a product strategy is a must-have skill for every product manager.

Whether you're a brand new or experienced PM, you've likely noticed "product strategy" gets used ambiguously. Truthfully, product strategy represents something different everywhere you go. There's no commonly-accepted definition, so probe on the meaning when you hear this term. (If this confused you before, breathe easy now... you weren't going crazy!)

At its core, a good product strategy represents a set of decisions about a product that the team will pursue to accomplish a business outcome and succeed in the market. Confusion arises because companies differ over which sets of decisions to include in the strategic document. Some companies might expect to see decisions about target markets, value propositions, and features or capabilities to build, while others may also want business model details or commercialization plans.

When crafting a product strategy, ask your team for a previous example to see what partners expect. But beware! Many companies have mediocre strategy documents (or worse nonexistent strategy documents). In situations like this, you'll need to take the reins. If you don't have a great example or need inspiration to take it to the next level, then try this method.

Use When

You need to finalize and communicate a set of key decisions that will align partner teams, give your company a strategic advantage, and guide the roadmap.

Average Time

1–3 business days, plus time for feedback and iteration

Try This

1. **Start by declaring the business objectives and vision for the product.**

   A good product strategy represents the connective tissue between high-level company objectives and tactical items listed on the product roadmap. It should immediately help the audience connect dots between why (business goals), what (product vision), and how (the areas of immediate focus for the team) without going into every specific detail.

   Presenting a product strategy without context will confuse your stakeholders. Chances are leaders won't have deep knowledge about a particular goal, problem, or solution, so they'll need your help switching context. To help, start your strategic document with a few brief reminders before zooming into the details:

   - Purpose of the document
   - Business objectives addressed by this strategy
   - Target markets included in the strategy
   - Product vision for the market

PRO TIP **KEEP THE VISION, STRATEGY, AND ROADMAP DISTINCT, BUT TOGETHER**

Some companies confuse product vision for product strategy, while others conflate strategy with the roadmap. In this book, we're keeping all three separate. If you work at a company that conflates these, be sure to always show the product vision, product strategy, and roadmap together. Combined, these represent your product plan.

2. **Identify 3–5 strategic initiatives that will accomplish the vision and provide focus.**

   A product strategy should represent 3–5 key decisions that will accomplish the product vision, guide the team's work, and generate unique sources of power or value for your product. When done well, it provides focus for the team and company resources. Some companies refer to these decisions as strategic pillars, bets, themes, or initiatives. Others might refer to these as objectives and key results (or *OKRs* for short). For this exercise we'll use the term *strategic initiatives*, but regardless of their label, these should represent concise, compelling opportunities where your team will concentrate resources, time, and money to win customers, gain a strategic advantage, and accomplish the vision. For example, strategic initiatives could represent major bodies of work needed to:

- Deliver the value proposition to users, including features or key benefits you plan to offer.
- Differentiate yourself among competitors, including new critical capabilities required to deliver a competitive advantage in your experience or value chain.
- Accomplish specific product goals, such as growth goals, revenue goals, or profitability goals.
- Validate critical hypotheses that require in-market feedback, such as pricing, positioning, or distribution hypotheses.
- Expand to a new market, by adapting or changing parts of an earlier product strategy.

If you find yourself overthinking it, consider where you are today (current state) and where you want to be (target state). Then write down what must be true to arrive at the target state. To help break it down, ask yourself these questions:

- What outcomes should the team pursue in the next 1–2 years to accomplish the business goals and product vision?
- What major bodies of work must we address to accomplish these outcomes, and which ones are unique to our company?
- Can we pursue these outcomes in parallel, or do they need to happen in a specific sequence based on our existing capabilities and estimated resources?
- Of everything listed, what are the 3–5 most important decisions we should activate *right now to gain a competitive advantage*? (That last part is critical. You might have your eye on next steps, but a product strategy needs to represent what the team is doing *right now* to work toward those goals.)

3. **Draft concise statements and structure your thinking.**

   After you've identified 3–5 strategic initiatives, try drafting them in concise language. You can list these initiatives as stand-alone phrases that reflect an outcome and approach, or you can include goals with them, as shown in Table 7.3.

TABLE 7.3 EXAMPLES OF STRATEGIC INITIATIVES

Product Description	A company entering a brand-new market with a new product	A B2B product focused on growth goals	An established B2C subscription product focused on revenue gains
Strategic Initiative	Validate product-market fit and fine-tune revenue model by optimizing our product and pricing strategies using early market signals and customer feedback.	Drive growth by offering the #1 integration experience for enterprise customers.	Increase monthly recurring revenue by offering personalized recommendations that engage and retain existing customers throughout their onboarding, service, and renewal journeys.
Goals	Convert 10% of qualified leads and retain 50% of them after six months.	Drive growth by integrating with 20 new enterprise customers this year.	Grow monthly recurring revenue by increasing customer retention 10pp this year.

Your strategy format often depends on company culture and preferences. Some will expect slides, visuals, or long-form documents. Pick whatever works best for your stakeholders but make sure that your product strategy doesn't exist in a vacuum. Said differently, the document must represent integrated, laddered thinking that connects the company's objective, product vision, strategic initiatives, and roadmap focus areas (see Figure 7.3 as an example). Before you finalize the document, it's helpful to assemble the puzzle pieces shown in Figure 7.3 into one view so you can make sure the language and decisions connect from top to bottom.

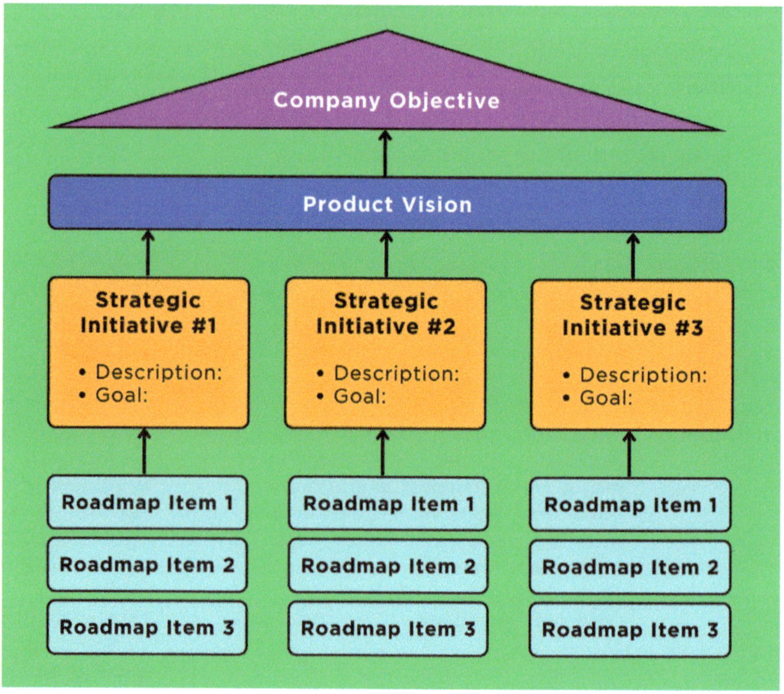

FIGURE 7.3
The vision, strategy, and roadmap should reinforce each other in support of company objectives.

4. **Pressure-test the draft product strategy.**

   I feel obligated to repeat myself here: many, many strategies are underwhelming. Don't fall prey to the mediocre strategy club. As a PM, you are paid for the ability to make good, strategic decisions. Crafting a product strategy is not a weekly, monthly, or even quarterly task, so when you get the chance to define one, put in the effort and use this as an opportunity to prove your worth. When you've completed the product strategy draft, use the checklist in Table 7.4 to determine if your product strategy passes the sniff test or if it needs more TLC.

5. **Socialize with experts and iterate.**

   If the product strategy is ready for prime time, ask trusted advisors for feedback first. Whereas product visions are meant to inspire and compel, product strategies are meant to drive realistic decisions and return-on-investment conversations, so find reviewers who think like P&L owners or executives.

As you receive feedback, fine-tune the product strategy and address blind spots. That said, it's nearly impossible to please everyone (and of course, everyone has a different POV on what a strategy should include). So, do your best to integrate feedback early and consider adding an FAQ section for your product once you start sharing with leaders and executives.

TABLE 7.4 THE PRODUCT STRATEGY SNIFF TEST

The Bad Product Strategy Checklist	The Good Product Strategy Checklist
☐ Looks like a set of features.	☐ Looks like a set of critical decisions and investments for our product that reinforce each other.
☐ Includes generic no-brainers like "make money," "be fast," or "acquire customers" without specific details.	☐ Offers a strategic point-of-view on how the product will create tangible value and gain an advantage over competitors and alternatives.
☐ Describes a product that tries to be everything to everyone.	☐ Represents a targeted set of focus areas, including customers/personas or needs to solve.
☐ Cannot help teams easily prioritize feature ideas or make trade-off decisions.	☐ Helps the team prioritize if new feature ideas should be considered.
☐ Doesn't show teams how their daily work connects to a specific vision or company goals.	☐ Provides a way for teams to trace roadmap items to key themes, the product vision, and company goals.

LEARN MORE

STRATEGIC FRAMEWORKS FOR PRODUCT THINKERS

Many product teams leverage frameworks during strategic thinking exercises. Check out some of the popular ones below and consider folding these into your team's practice.

- Business Model Canvas by Strategyzer
- Porter's Five Forces by Michael Porter
- The Four Ps of Marketing (price, product, place, and promotion)
- SWOT Analysis (strengths, weaknesses, opportunities, and threats)
- Objectives and Key Results from the book *Radical Focus* by Christina Wodtke
- Product Strategy Canvas by Paweł Huryn

PLAY 14

Building a Roadmap

What will we launch, when will we launch it, and what impact will it have?

Product roadmaps communicate priorities and implementation goals over specific timeframes. They serve three audiences: teammates who need goals, priorities, and delivery timeframes; leaders planning resources and coordinating cross-team dependencies; and partner teams like operations, marketing, or sales who need visibility to adapt their plans.

Roadmaps are easier to create than product strategies because they represent tangible, concrete work the team will pursue over specific timeframes. That said, PMs adjust roadmaps more frequently than a vision or strategy as teams rearrange work or change timelines. Never consider the roadmap "final"—it's a living document that represents a point-in-time view. (Keep reminding stakeholders of this, too!)

Like strategies, there's no single way to visualize a roadmap—it largely depends on what your stakeholders need to see and how you want to communicate goals and progress. You'll probably end up using different templates or creating unique views based on your stakeholder audience, so be prepared to stay flexible. That said, here are common steps you can take to build a roadmap that works.

Use When

You need to form and communicate an implementation plan that supports the product vision and product strategy.

Average Time

2–3 days to build the initial roadmap, plus time for feedback and future adjustments

Try This

1. **Select the roadmap type that's right for your audience.**

   When viewing the roadmap, most teams want to understand which work the team will accomplish and by when. The trick is to determine the right level of detail for both items and identify if your stakeholders need to see anything else, such as goals, metrics, or milestones. To get started, consider the types of roadmap elements in Table 7.5 and partner with your leader and stakeholders to prioritize the right level of information.

TABLE 7.5 COMMON ROADMAP ELEMENTS

Work Description	Time Horizon	Additional Context
Strategic initiatives	Now, next, later	Goals/metrics
Epics/features	By quarter, month, week	Themes
Hypotheses to explore	By release or milestone	Dependencies

2. **Prepare the roadmap scaffold.**

   After you select roadmap elements to include, build a scaffold or template for your roadmap. Most PMs start by listing the time horizon over the top of the document and themes down the left side before layering in details (as shown in Figure 7.4). And no need to reinvent the wheel here—grab shamelessly from existing templates your team has or pull directly from the web. There are hundreds of roadmap templates out there.

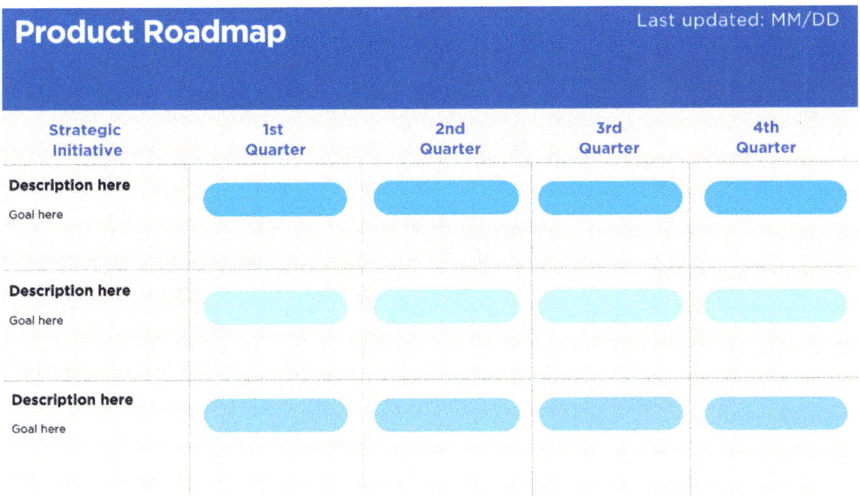

FIGURE 7.4
Use this figure as a guide and adapt it based on your strategic framework and time horizon choices.

3. **Put in the work (literally!).**

   The first two steps were easy, but this part takes real brainpower. Map out all the known work you must accomplish over the roadmap timeframe and sequence it on the roadmap (as shown in Figure 7.5). This step might require you to:

   - Schedule a multihour collaborative session with your business, design, and tech counterparts.
   - Synthesize items from the Problem-Solution Fit Map, technical feasibility assessment, and product strategy artifacts.
   - Decompose work into discrete efforts or "features."
   - Collect time-bound estimates from feasibility experts.
   - Identify maintenance work or must-do compliance work.
   - Complete a sense-check with the team to determine if the roadmap looks reasonable based on the scope, resources, and estimates available.

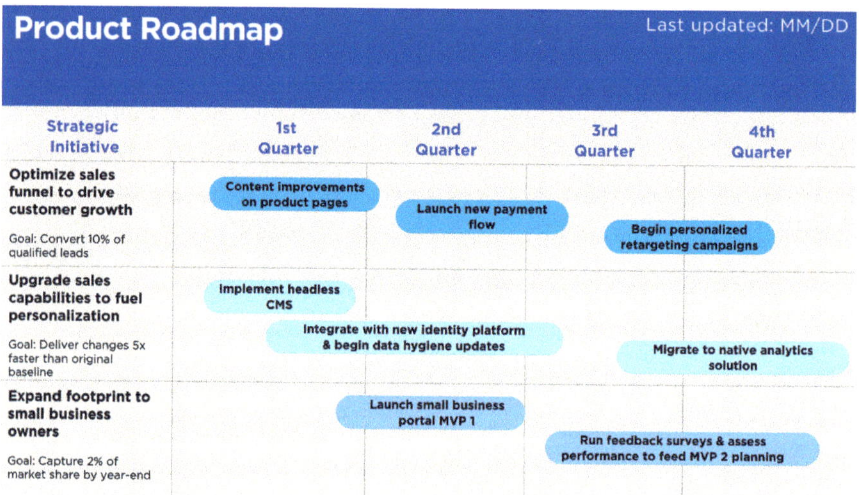

FIGURE 7.5

Post your complete roadmap in a shareable location so any stakeholder can see the latest version.

4. **Use colors/visuals to communicate different themes.**

   Consider using colors or visuals to represent additional context your stakeholders need to see. For example, should you highlight work with major dependencies? Do stakeholders need to easily spot work that drives customer-facing change versus must-do work to maintain the platform or stay compliant? If these items aren't easy to spot, consider color-coding or visual changes to make it clear.

5. **Keep it updated and visible for everyone to see.**

   After you've finished the roadmap, throw a "last updated" date on it and start shopping it around. And yes, put it in a location everyone can access without needing to ping you.

The product roadmap is perhaps the most popular artifact you'll create. Use it during product pitches, strategic planning sessions, onboarding conversations, and of course, stakeholder updates. And remember, don't treat it as *final*—treat it as *current*.

Establishing Sponsors and Durable Support

Do I have support to pursue this?

There's a good reason why every job description for product managers requires the ability to "influence without authority." You can have the world's smartest product concept, and your product could still fail without durable company sponsorship and resources to deliver or maintain the solution. So, how can PMs influence key decision-makers to pursue a new product strategy, solve an unaddressed need, or even build in time for user research? Start by understanding your company's operating model, planning approach, and key decision-makers.

Know Your Operating Model

A company's operating model represents exactly how the company selects, funds, and manages strategic priorities. Understanding your company's model helps you stay one step ahead of stakeholder questions. Operating models and planning methods differ by company size, industry, and organizational maturity, but here are a few common models.

Product Operating Model

Popularized by Silicon Valley Product Group, this model emphasizes continuous value delivery by establishing empowered teams focused on *outcomes* rather than feature lists. Typically, product, design, and technology leaders review company strategy, build the product vision and strategy, and then fund specific outcomes during annual or semi-annual planning events. Once funded, outcomes are assigned to cross-functional product teams with monthly or quarterly check-ins.

In this model, PMs have autonomy to self-fund new solutions within their defined problem area, but must make the work transparent, show expected impact, and highlight trade-offs during routine checkpoints. If the PM recommends a strategic pivot that requires major roadmap or resource changes, they often present a business case including problem statements, product vision, strategy, metrics plan, and roadmap.

Project-Based Model

In this model, business unit leaders (not product, design, or technology leaders) prioritize initiatives and decide funding decisions during strategic planning events. Once a project is approved and funded, the company establishes short-term project teams to build and launch solutions for a finite period. These organizations may suffer from tactical, "shiny object syndrome" and under-invest in existing solutions. Success often requires strong relationships with business leaders and a plan that generates maximum value with finite resources.

A Company Transforming to a Product Operating Model

Many companies are caught between legacy project-based approaches and evolving product models. If that sounds like your environment, focus on building and maintaining trust with business sponsors while navigating a transformation curve with your product leaders. Sometimes, the best way to build trust with business leaders is to focus on quick, tactical wins first to show visible progress. But while you're knocking out tactical wins, start product discovery efforts and share research insights with everyone. Once you have enough insights, use collaborative workshops to build a vision, strategy, and roadmap with your partners and then work with product leadership to secure resources. Be prepared to practice organizational patience in this environment—transformation takes time.

Special Projects Model

Many companies set aside extra funding to form innovation groups or special project teams. In some cases, the company will host hackathons or pitch competitions where individual contributors (like you) can propose new concepts and compete for funding. Projects in this model tend to be highly experimental, so PMs must be prepared to share concise problem statements, hypothesis-driven learning plans, and creative ways to experiment. In some cases, the company may follow a stage-gate approach and require teams to present key learnings and product concept details, including desirability, viability, and feasibility analysis before they can secure additional funding.

Anything Goes

In this scenario, the company doesn't have a defined planning, prioritization or funding process. PMs should light a candle and pray for the best! In all seriousness, when the company lacks portfolio management practices, decisions tend to be highly political. PMs will need to focus heavily on understanding key decision-makers and building relationships with them to navigate this environment.

Think Roadshow, Not Circus

Securing executive sponsorship is a major win, but what happens when your plan relies on other teams that answer to *different* executives? This happens every day in companies with numerous executives, strategies, products, platforms, and competing demands. In product speak, it's referred to as *dependency management.*

Dependency management requires strength in numbers and alignment at several levels of a company, so PMs should inform their leaders and leverage dependency processes to socialize requests across silos as early as possible. And while support from your leaders can help, completely off-loading dependency management to others isn't a good idea. As a PM, you have the best context on the problem *and* solution details, so translate this knowledge into product roadshows to share context, request support, and rally the troops.

A product roadshow is a meeting you host to evangelize your team's work, build alignment with stakeholders, and vocalize a request for support. The meetings are usually brief (30 minutes) and include a short, reusable presentation (think 3–5 slides) that you develop to share essential context about your initiative before asking for support. By creating a reusable presentation, you can limit the amount of rework on yourself and follow a rinse-and-repeat pattern each time you need to build bridges with new partners and stakeholders. Figure 7.6 shows the typical outline for product roadshow content.

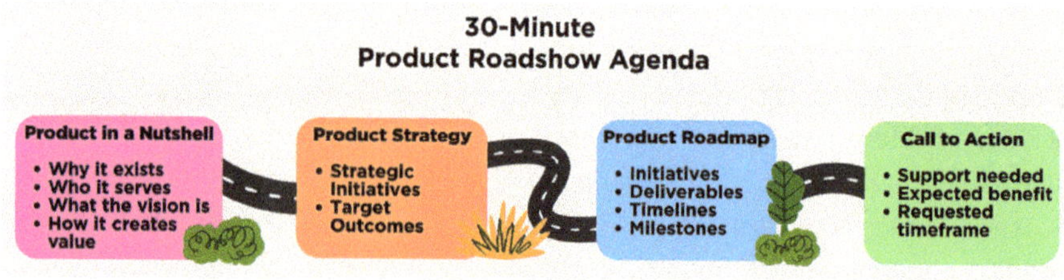

FIGURE 7.6

You can build this once and adjust the final call to action based on your audience.

Product roadshow slides support other use cases, too. For example, these materials come in handy when you need to:

- Present a pitch during a hackathon, pitch competition, or strategic planning event.
- Onboard new teammates to your product.
- Remind stakeholders about the original goals during progress checkpoints (to keep the team honest).
- Build adoption for your product, tool, or service across the company.
- Submit your team's work for an internal award.
- Complete your annual review or present a case for promotion.

Timing is also an important part of dependency management. Your roadshow slides won't impress partners if you're asking for support so late that it would impact their other commitments. Engage your dependent partners as early as possible to prevent chaotic, last-minute requests and maintain a good brand. This gives your partners time to plan and rework their own priorities if needed.

PRO TIP **MANAGING A COMPLEX PROJECT? BUILD A STAKEHOLDER MAP.**

Stakeholder mapping is the process of identifying and visualizing all the individuals that have a stake in your product's success, understanding critical influencers in the group, and clarifying what matters to partners. A stakeholder map can help you identify the most important relationships to prioritize at different stages in the product development process, increase decision velocity by understanding roles and responsibilities, and improve communication with partners to reduce unwanted surprises. There are tons of templates and examples online. Check out examples from Miro and Figma as a starting point.

Pulling It All Together

This chapter provides a lot of methods that produce documentation, but keep in mind that the product manager's job isn't to build documentation. Your job is to develop a point of view based on insights, align the team around a solution that will create value for the customers and company, and back it up with a plan that will help teams accomplish it. And you must do this without direct authority over many functions and partners.

While it's possible to accomplish this level of strategic thinking and to align partners through conversation, artifacts like a metrics plan, product vision, product strategy, and roadmap help crystallize your point-of-view and give stakeholders the context to start supporting you. Once you've got alignment and resources to pursue a solution, get ready to shift gears out of discovery and into something that requires a whole new set of skills and methods—product delivery.

EXPERT ADVICE WITH CHETAN KANDHARI

WHAT C-LEVEL EXECUTIVES WANT FROM PRODUCT MANAGERS

As the Chief Innovation and Digital Officer of Nationwide, Chetan Kandhari leads a dynamic innovation team and product portfolio that spans B2C, B2B, and B2B2C solutions. Here's the must-know advice he gives to product managers who want to build a strong brand with executives and lead innovation work.

Corporate innovation functions typically follow one of three models to fuel breakthrough impacts for an established company. One model is standing up unicorn innovation teams focused on finding the next billion-dollar idea. These teams typically exist with high autonomy, build their own culture, and don't engage with core company practices. The second model is creating enterprise innovation teams to solve adjacent needs for the company's current markets, with the intent of transferring new solutions to the core business if proven successful.

continues

EXPERT ADVICE WITH CHETAN KANDHARI (continued)

The third model is establishing business unit innovation teams that focus on reimagining value chain improvements or creating scalable enterprise assets that support multiple business lines. For example, Uber's app is a scalable asset that supports distinct revenue streams: a ride hailing service and a food delivery service.

If you're a product manager supporting any of these models, you must become an expert at pitching new solutions to executives and investors responsible for funding breakthrough change. The best pitches are concise and pack a punch. They distill critical insights about the opportunity and articulate the value proposition for a new solution in three sentences or less. They translate strategic concepts into immediate next steps with an aggressive roadmap that shows how you plan to accomplish key milestones with funding and support. And finally, they include critical assumptions, so stakeholders understand risks and rationale before agreeing to fund a new solution.

Innovation product managers should also know that success isn't just about solving customer needs. Innovation efforts often require significant up-front investment with a longer road to payoff, so you need to build sustainable corporate sponsorship and political buy-in across the organization. The best product managers do this by building bridges with other departments, recognizing partner teams, and working to keep alignment with their leaders and peers in every direction.

And finally, you need to demonstrate skills and passion for the work. On paper, the best product managers excel at problem-solving and unlocking expertise from their cross-functional teams. They index heavily on hypothesis-driven product development and spot signals that help determine if new products are on track or need to pivot. However, innovation is about dealing with the unknown and ambiguity, so I also look for product managers who demonstrate passion, ownership, and curiosity. These traits cannot be overstated because they power teams through ambiguity. Those who expect a straight line toward product-market fit or get caught in analysis paralysis aren't cut out for this type of work. You need to harness passion and drive.

CHAPTER 8

Building Products

Who sounds like a stronger candidate during a PM interview—someone who has only worked on product strategy or someone who has shipped and managed products in-market?

(It's the latter.)

While some PMs love cooking up big ideas, a product strategist without execution skills is just a thinker. To be an exceptional product manager, you can't just be a discovery expert. Companies need thinkers *and* do-ers. In fact, many hiring managers favor candidates with delivery experience over discovery chops for individual contributor roles because forming strategy usually happens once or twice a year while continuous delivery and in-market optimization happen daily.

Realistically, most PMs join companies that have existing products in-market and a backlog of business opportunities waiting for resources and attention, so hiring managers are looking for dynamic team players who can immediately start managing existing solutions while also leading discovery and development for new opportunities in the pipeline. And can you guess when they want you to accomplish all of this? By yesterday!

Some product managers shy away from delivery and prefer to focus on customer research, brainstorming ideas, and dreaming up strategy because it's creative and fun work. But the truth is, there's no better proof of an exceptional product manager than someone who can form a vision and strategy *and* can bring it to life. You absolutely, unequivocally must know how to develop and ship products if you want to be successful in this field. Trust me, you don't want to be the person that "does strategy" but can't back it up.

Delivery is not an exercise in throwing a roadmap, features, or requirements over the wall. Customers need someone who can represent their needs and discourage quality shortcuts when teams build solutions. Engineering teams need a partner who can make smart, defensible trade-offs based on customer and business impact when feasibility constraints get real. And businesses need someone who can ensure that the final product accomplishes the original objectives. Given these desires, who is the person with the *absolute best context* to guide a delivery team?

The product manager.

Truthfully, there's nothing better than in-market feedback to know if you're actually a good product manager. And to get there, you have to lean in and master product delivery. Thankfully, there are loads of experts, methodologies, templates, and life hacks to help you navigate execution with grace and grit, and this chapter covers the most essential ones. Whether you're driving a brand-new product into market or managing an existing solution, these are the questions you'll need to answer when it's time to deliver on the roadmap:

- What is my responsibility during delivery?
- What does the team need to be successful?
- How do I break down the roadmap into work?
- How can I validate that the solution meets expectations before launching?

Shifting to Delivery Mode

What is my responsibility during delivery?

Product managers are still responsible for representing the voice of the customer and business during delivery, but instead of evaluating insights and forming hypotheses about value propositions and feature sets, they are decomposing user requirements with the team, assessing daily progress, and overcoming risks that jeopardize product quality and launches. To do this well, you must learn to master different levels of zoom. In one meeting, you'll zoom in to discuss functionality details for a specific user scenario, and in the next meeting, you might have to zoom out and discuss if the overall product quality and team velocity is on track to meet business goals.

And that's just the tip of the iceberg. During delivery, product managers are also responsible for overcoming indecision, avoiding endless discovery, preventing development slowdowns, resolving dependency conflicts, making trade-offs, protecting the team from scope creep, and navigating bureaucratic processes. In many cases, leaders expect their teams to practice extreme ownership during execution, which means anything that impacts the team's ability to deliver the product in a timely manner becomes the PM's responsibility. Table 8.1 highlights the most common tasks PMs support during product delivery, and in many instances, you're doing them while leading a separate discovery effort at the same time.

TABLE 8.1 THE LIFE OF A PM DURING EXECUTION

PM's Responsibility	Everyday Tasks
1. Provide clarity. 2. Ensure quality. 3. Enable efficiency. 4. Maximize value.	• Articulate company goals, product vision, and product strategy. • Define and communicate customer and business requirements. • Plan work and decompose with your team. • Prioritize work in team backlog. • Make trade-off decisions. • Support product development tasks, including testing and releases. • Identify and unblock dependencies. • Mitigate risks to quality, budget, or timeline. • Monitor delivery progress and team metrics. • Communicate updates to stakeholders. • Adapt product roadmap and backlog based on market, user, or research insights. • Track product performance and key metrics post-launch.

During large or complex product rollouts, some companies staff specialized roles to keep work on track, such as program managers, dependency managers, Scrum Masters, or product operations experts. But in most cases, these specialists support multiple teams at once, so you can't wait for them to rescue you. To ensure that your team is on track and development velocity is healthy, you'll need to embrace and practice skills that draw from project management and Agile product owner roles, many of which are already represented in Table 8.1.

Speaking of product owner roles, let's address an important topic. Many product managers try to separate themselves from the "product owner" label because it's often associated with feature factories (companies that expect PMs to constantly deliver new output based on customer or stakeholder requests with little consideration for product-market fit or quality). And while every PM needs to demonstrate strategic, big-picture thinking, exceptional PMs know when to put their product owner hat on to effectively bridge strategy to execution. Instead of being stuffy about titles, focus on the skills. And if you haven't considered formal product owner training or learning about this role, read up. The Scrum Alliance offers training and certifications that can help.

Establishing Ways of Working

What does the team need to be successful?

In many cases, PMs join existing small teams (sometimes referred to as *pods, squads, sprint teams,* or simply *Agile teams*) with established roles, processes, and behaviors across design, technology and data partners. Ideally, the PM simply fits right in like a puzzle piece because roles and responsibilities are clear and understood by all. But if you are forming a new cross-functional team from scratch or you're unsure how the current team functions, start by clarifying ways of working for the team.

MY TEAM IS UNDERPERFORMING... WHAT DO I DO?

Subpar products, role confusion, communication breakdowns, and consistently missing milestones are a few signs that your team might be underperforming. And when the team is underperforming, it's your job to fix it. (Remember the part about extreme ownership? Helping your team get unstuck is one way to practice it.)

If you notice team confusion, conflict, or underperformance, it's time for a heart-to-heart chat with peers about what's working and what needs improvement. Start by bringing the business, tech, design, and product leads together and completing the audit in Table 8.2. As a group, identify and prioritize the most critical issues to improve (i.e., team goals, requirements readiness, development velocity, role confusion, staffing needs, etc.) and co-create solutions together. These can be new guidelines, a team charter, a request for additional head count, or even small adjustments to existing processes. For inspiration, research concepts like Dual-Track Agile, Scrumban, 2-pizza teams, and other team practices popularized by the top tech companies.

Once you're ready to roll out improvements, bring the entire cross-functional team together with leaders, share revised expectations for ways of working, and provide a target date to start practicing new behaviors. After instituting changes, observe what's working or not working for one to two months, gather feedback from the team, and update agreements as needed. Stay patient, open-minded, and provide training when necessary—transformation takes time, repetition, and encouragement.

Ways of working is a concept that represents the behaviors, processes, and operating model a team follows to accomplish their goals. To understand or define ways of working on a team, consider the 3 P's: people, process, and product. Being intentional about each of these categories is critical in establishing healthy, high-performing teams. Table 8.2 provides an easy guide to audit and clarify the ways of working on a cross-functional team.

TABLE 8.2 TEAM WAYS OF WORKING AUDIT

Theme	Clarifying Questions
Product	• What is the primary objective for this team? • What part of the product vision or product roadmap do they support? • What are the team's key metrics or targets? • What are their immediate priorities for this product? • What's in the team's product backlog?
People	• Who are the people on the team? • What skills do we need to be effective? • What are their roles and responsibilities? • What specific deliverables does each person own? • Does everyone understand roles and responsibilities across the team? • Where are there gaps in skills or support?
Process	• How does the team prioritize work and make trade-off decisions? • What software development process does the team follow? • What methodology does the team follow to plan, track and accomplish work? (For example, Scrum, Kanban, Dual-Track Agile, Scrumban, etc.) • What project management software does the team use? • What ceremonies, meetings, or communication outlets does the team use? (For example, standup, sprint planning, refinement, retros, design critiques, etc.) • What standards, commitments, or agreements must they follow? (For example, Definition of Done, Enterprise NFRs, etc.) • How does the team manage dependencies and risks? • How does the team perform testing? • How and when does the team release product updates? • How does the team monitor and report on metrics? • How does the team communicate progress to stakeholders?

Hosting Team Ceremonies

By this point, it's obvious that delivery isn't just about coding software products and managing product releases. Of course, you need the technical knowledge about lower environments, testing frameworks, and release processes to understand progress. But mastering delivery requires impeccable cross-functional team leadership skills to actually help the team across the entire development life cycle. To be effective, PMs must operate like a player and a coach at the same time.

Managing your own workload while supporting the team can get tricky if the team doesn't have a bit of structure and clear guidelines around communication. While you shouldn't go overboard with too much structure or too many meetings, it's important for PMs to host and attend agenda-driven, intentional checkpoints with the entire team. Messaging tools are fantastic, but you can't solve everything over Slack. Figure 8.1 highlights the essential team ceremonies that every PM should schedule and attend on a recurring basis. Having these on the calendar gives the team a reliable outlet to discuss their current work, escalate issues to you, and help you refine upcoming work.

THE TEAM CEREMONY STARTER KIT

CEREMONY	FREQUENCY	PURPOSE
Planning	Bi-weekly (45 min)	• Discuss team capacity over specific timeframe. • Pull items backlog into current iteration until capacity is achieved. • Assign work to specific teammates.
Backlog refinement	Twice a week (60 min)	• Introduce features or requirements several weeks before development should begin. • Refine and decompose them into smaller units of work. • Update level of effort estimates. • Identify dependencies and steps to manage them.
Standup	Daily (15 min)	• Discuss daily progress on items in current iteration. • Raise issues or blockers and determine steps to overcome.

FIGURE 8.1
Team ceremonies build a culture of trust and drive efficiency on the team.

PLAY 15

Building a Backlog

How do I break down the roadmap into work?

Establishing expectations and an operating model early helps clear the way for the team to get stuff done. And to get stuff done, the team needs a healthy backlog of work.

A product backlog is a visible "to-do" list of work the team must complete during product development, and the product manager is responsible for building and updating it. Whereas product roadmaps offer a high-level view of work and timelines for a product, the backlog is more granular than a roadmap and guides the small, cross-functional team that designs, develops, and delivers software.

Having a prioritized backlog of work helps the team avoid velocity slowdowns. A good rule of thumb is to maintain at least 30 days of development-ready work for the team. Creating a 30-day backlog from scratch requires intensive collaboration with the team to define and polish product requirements. After building the initial backlog, you'll need to keep it updated with development-ready work as new opportunities emerge. The best way to maintain a healthy backlog is by working on requirements several weeks ahead of your design and development teams and hosting ongoing refinement sessions several weeks before development begins. Working ahead of development gives the team breathing room to close requirements gaps, manage dependencies earlier, and avoid quality shortcuts. Here's how to get the backlog started.

Use When

You need to break the roadmap down into achievable increments and tasks for the team.

Average Time

Each feature takes ~2 hours to refine and decompose with the team, after initial requirements are done

Try This

1. **Learn how your project management software tool works.**

   Most companies use project management software tools like Jira, Asana, Azure DevOps, or others to document, publish, and maintain their product backlogs. Before attempting to build a backlog, learn how your tool categorizes and classifies project work.

2. **Translate themes or opportunities from the roadmap into discrete features.**

   While tools differ in terminology, most PMs form backlogs that consist of *features* and *user stories*, as shown in Figure 8.2. Features represent a meaningful piece of functionality that provides concrete value to the end user when delivered. Features should ladder into strategic themes or initiatives (sometimes tracked as *epics* in workflow tools) and include strategic context and complete requirements. User stories represent smaller units of work required to deliver the feature. They help teams *decompose* features or larger efforts into smaller, achievable work items or tasks, and they typically describe the persona, need, and benefit written from the end-user's perspective for a specific scenario.

   Once you know how to use the project management software, review your roadmap and translate it into a set of discrete features in your company's project management software tool. Simply create a unique work item or "shell" for each discrete feature or capability.

FEATURE & USER STORY EXAMPLES

FEATURE	USER STORY
Implement digital experimentation service into sales app	• As a product manager, I need a way to determine customer eligibility for my experiment so I can deliver tailored experiences. • As a product manager, I need a way to validate our experiment is not doing harm after launch so I can minimize risk to customers. • As a product manager, I need a way to study test results in real-time so I can make timely decisions to stop or resume the test.
Implement comparison charts during sales flow	• As a customer, I need a way to understand product details and benefits, so I can easily compare between options. • As a customer, I need a way to easily understand payment methods and options, so I can quickly set up payments and confirm my purchase.

FIGURE 8.2
Here are some examples showing how to decompose features into user stories. (Although some companies follow the Jobs-to-Be-Done framework.)

3. **Add clear requirements to each feature.**

   For each discrete feature, start listing a clear set of functional and nonfunctional requirements in the description. Add reference materials and design specs as well. As a rule, the team should be able to understand the feature requirements without needing you to explain it. (The next method provides more depth on requirements writing.)

4. **Refine, decompose, and estimate features with the team.**

   When you're ready to share new feature requirements with the team, invite the team to a backlog grooming or refinement session. During the session, review the feature requirements together, allow the team to probe or ask questions, and then update your requirements to improve clarity.

   After the requirements are clear, start the process of decomposing the work with the team by breaking it into smaller user stories with clear acceptance criteria. During the session, the development team should determine the technical approach to deliver the work for each user story, call out any known dependencies, and estimate the level of effort for each item.

PRO TIP PRACTICE VERTICAL SLICING WHEN DECOMPOSING FEATURES

Vertical slicing is a concept where teams build user stories and acceptance criteria in a way that results in working, useful software when the story is completed. These stories usually require full-stack work, which helps the team spot and correct issues faster during integration. This approach combats horizontal slicing where the user story only reflects work needed in one layer of the tech stack but doesn't deliver working software once it's completed.

5. **Rinse and repeat for anything else in the backlog, including bugs and tech initiatives.**

   Backlogs should represent a single source of truth for the team's work. That means your backlog and refinement sessions should discuss and refine all work needed to build and launch new products, enhance existing features, resolve defects or bugs, and support initiatives that keep the product or platform resilient, similar to what you see in Figure 8.3. To do this well, engage your tech partner and operations teams to gather and document info proactively about upcoming tech maintenance work and critical production bugs the team must address.

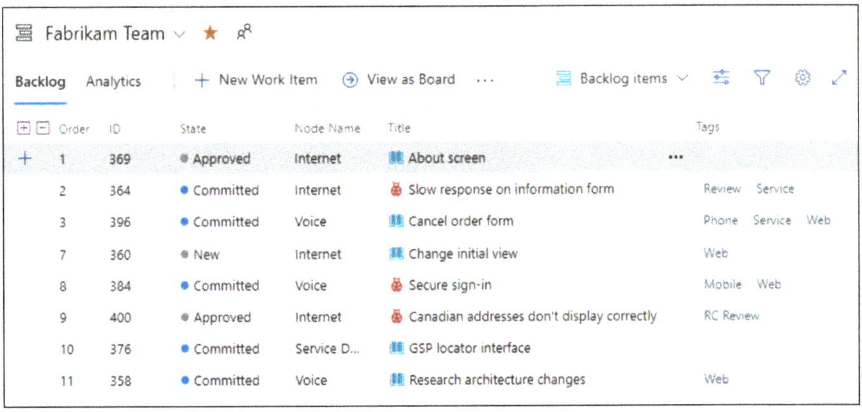

IMAGE COURTESY OF AZURE DEVOPS

FIGURE 8.3
Each team should have a single backlog to reflect all work in one location.

PLAY 16

Crafting Product Requirements

What does the team need to be successful?

So far, you've heard a lot about requirements—the different types of requirements, the importance of requirements, how requirements feed backlog development, etc. But you might be wondering how to actually write requirements or what "good" looks like.

Truthfully, it depends. Some companies practice a "less is more" ethos when it comes to documenting requirements because they don't want PMs writing a novel and slowing down progress. Other companies practice a "more is more" approach with 10-page requirements documents because they want to make sure every single detail is crystal clear and traceable, especially in regulated industries.

If your company doesn't have a clear approach, decide what's best for your team based on company culture and context. If cross-functional collaboration is breezy, product quality is high, and your team can resolve or enhance issues in production quickly, then you don't need overly detailed requirements. On the other hand, if you're working on an app with millions of customers, supported by dozens of teams, and the product involves significant business risk, then well-documented requirements are a useful way to improve cohesion and quality. Either way, the process of drafting requirements helps you crystallize your thoughts about a problem and influence stakeholders, so you need to know how to write them. (And yes, you can also leverage AI tools to help you turn this method into a reusable pattern!)

Use When

You need to translate specific roadmap features into clear guidance and acceptance criteria for the design and development team.

Average Time

Usually at least 2–3 hours to analyze and write initial requirements, plus extra time to refine with the team

Try This

1. **Start with strategic context.**

   All product (or feature) requirements must start by introducing the problem or opportunity in the clearest terms possible. In many cases, you'll need to share these with stakeholders or other teams, so do not use code names or acronyms or overexplain. Focus on the essentials and answer these questions:

   - What is the specific problem or opportunity you want to address? (Include data and details about the size, frequency, or severity.)
   - Who benefits from solving this problem? (Describe the customer segments and business areas that benefit from solving the issue.)
   - What is the expected impact of solving the problem or success criteria? (Include metrics, targets, and dollar impacts.)
   - Why is the problem worth solving right now? (Include market insights, competitor insights, and technology insights, if applicable.)
   - What's out of scope?

2. **Specify scenario-driven solution details.**

   The best requirements explain what must be true to consider the solution complete and successful. You can include business requirements, functional requirements, and nonfunctional requirements for this part. The key is to include discrete scenarios that will occur and articulate success for each one. If you're working on an experience or product that has many different target users, products, channels, and business rules, you may need to list many scenarios. Try using the formats in Table 8.3 when documenting each scenario. And don't just include the *happy path* customer scenarios, you need to specify error scenarios or *unhappy path* customer journeys, too!

TABLE 8.3 SCENARIO-DRIVEN REQUIREMENT FORMATS

	User Job Story	Behavior-Driven Development Scenario
Format	When _[this situation]_, I want to _[motivation]_, so I can _[expected outcome]__.	• GIVEN [this context for this target user] • WHEN [this scenario occurs] • THEN [the system or product should behave in this way].
Example	When I return a product and call in to speak with customer support, I want the customer service team to provide an accurate update about my order status, so I know if my order was received and the company will issue a refund.	• GIVEN a customer contacts us about a refund • WHEN the IVR system transfers the customer to our service reps • THEN the service rep's system should display a real-time summary of the customer's history, including contact information, order numbers, order status, products purchased, return status, and refund status.

Before you move on and call it a day, don't forget about including those nonfunctional requirements, too. We covered these in Chapter 5, "What: Defining Solutions" (see Table 5.3), but as a reminder, each feature or product must include requirements for the following:

- Performance, including availability, scalability, and system response times
- Observability, including the specific data the system(s) must collect, process, store, and display for reporting purposes
- Security, privacy, compliance, regulatory, or legal rules

- Accessibility, security, or other critical standards
- Testability, including prelaunch testing criteria (i.e., user testing, acceptance testing, etc.)
- Recoverability, including system recovery expectations if/when something goes wrong

3. **Host sessions to elicit, facilitate, and gather more information to refine.**

   Sometimes, you'll need to *elicit* requirements from subject matter experts and stakeholders that represent the needs of other departments, such as compliance, pricing, marketing, or other key stakeholder groups. Before you schedule a session, attempt to draft at least 50–75% of the entire requirements document prior to hosting a meeting, so the session is spent solving the hardest questions only.

   During the actual requirements elicitation session, start by introducing the product or feature, call out the specific questions or gaps you still need to answer, and then begin asking questions for the group to answer. As the group discusses options, screenshare the requirements and actively update them during the meeting. Before ending the session, ask the group to review the changes and identify any missing scenarios to refine. (By the way, consider recording the discussion. It's hard to hear everything the group says while you're writing, facilitating, and keeping track of time, so you may want to listen again later.)

4. **Finalize requirements, align on technical approach, and decompose with the team.**

   Eliciting and defining requirements is a process that requires fine-tuning until you've clarified the most critical elements, shown in Figure 8.4. Sometimes, it can take a few weeks to get it right, but once you've captured the voice of the business and customer intent, it's time to share them with your design and development partners during refinement or backlog grooming. The team may request a few tweaks or additions, but if not, start discussing the solution's design, technical approach, and dependencies. Once those are clear, you can decompose into user stories.

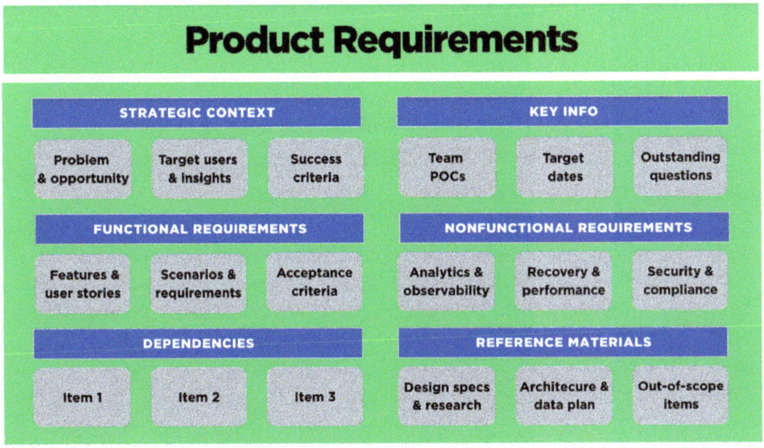

FIGURE 8.4
Product requirements should crystallize insights and decisions that provide context and clear direction for the team.

The Art of Prioritization and Planning

Building a backlog and writing awesome requirements are fundamental parts of your role during delivery, but how exactly does the team know what to work on and when? And what happens if something new hits your desk that's clearly more important than some of the items in your current backlog? Alas, you'll need to practice the ongoing art of prioritization and planning. And yes, I do mean art, because it's not exactly scientific.

Prioritizing the Backlog

The product backlog is not a static list of tasks. New features, bugs, or requests can enter the team's roadmap or backlog at any point, based on emerging priorities or new insights. When this happens, the team expects you to prioritize what they should tackle. Some companies have predefined prioritization frameworks to help PMs navigate these decisions. Many of them draw from popular frameworks, shown in Figure 8.5.[1] If your company doesn't have a clear

1 Don Reinertsen popularized Weighted Shortest Job First in his book *The Principles of Product Development Flow: Second Generation Lean Product Development.* Dai Clegg developed the MoSCoW prioritization method while working at Oracle, and Sean McBride developed RICE while working at Intercom.

prioritization framework, consider integrating one of these into your own practice with the team.

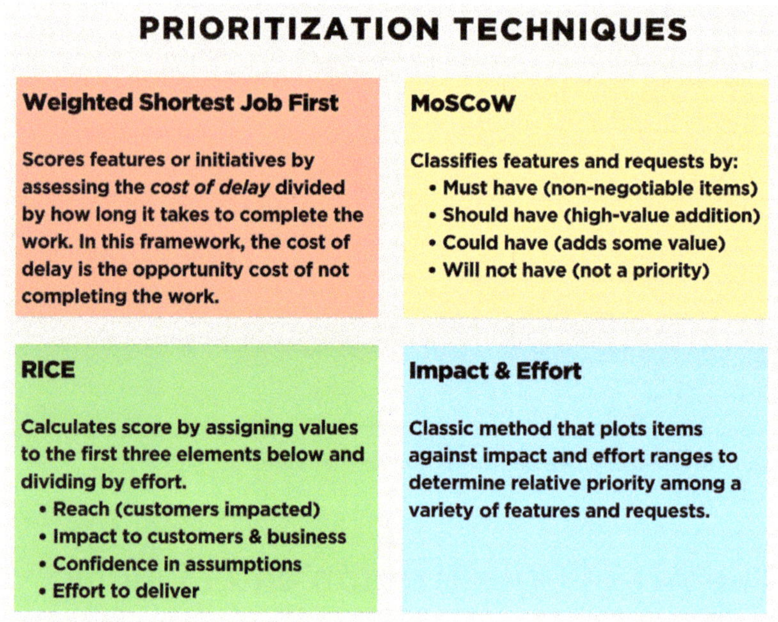

FIGURE 8.5

Partner with your team and stakeholders to align on the specific criteria the team will use to prioritize backlog items.

That said, I've yet to meet a PM who uses a prioritization framework as the complete gospel for every single decision when prioritizing work. Of course, every PM says they should prioritize the highest-value items, but sometimes prioritization frameworks don't account for real-life considerations like internal politics, handoff timing, or emerging risks. In some cases, the highest-value backlog item might not be fully ready for development, and you don't want the team waiting around, so you prioritize a few quick-win items for a few weeks. Or you might notice that tech debt is starting to accrue at an unsustainable rate, so you prioritize bug squashing and quality improvements for an entire sprint. Whatever you decide and whatever prioritization method you choose, the trick is to be consistent, be transparent, and ensure that stakeholders can easily follow your rationale when they ask why you chose X over Y.

Forming a Time-Bound Plan

Prioritization and planning go hand-in-hand. When you're leading a team that manages a product in-market, it's inevitable that new issues will emerge and trigger backlog priority discussions. To help teams stay focused, you need to balance the art of making commitments to delivering specific work while maintaining a bit of flexibility if new, more important priorities arise.

To manage this balance, many product teams embrace time-bound planning cycles where teams commit to completing a specific set of work items over a given timeframe or iteration. For example, teams that follow Scrum may adhere to a planning cycle where they commit to work in quarterly increments, six-week increments, or two-week sprints. Doing this helps the team lock in priorities for a finite period of time and stay focused without committing to work that is multiple months or years away. You can easily build a time-bound plan for the team in a few simple steps:

1. Determine the timeframe for your team's plan.
2. Then gather the team's total capacity (usually represented in days or hours) for the timeframe.
3. Next, review the individual estimates for features and user stories in the team's prioritized backlog and begin adding items to the time-bound plan until the team's capacity is full.
4. Share this with your tech lead and design lead for feedback prior to finalizing with the team during your planning ceremony.

Even if you commit to work over a few weeks, it's still important to set aside capacity for unknowns and bug work, especially if the team supports an existing in-market solution while building new features or products at the same time. For example, the team may decide to consistently allocate 20% of their total capacity every iteration to tackle bugs, defects, or unknown work that emerges during the increment. If you do this, be sure to document and display this in the team's iteration plan, so the team's total capacity and assignments are clear to everyone. Figure 8.6 shows what a sample plan looks like for a team using Jira to manage sprints.

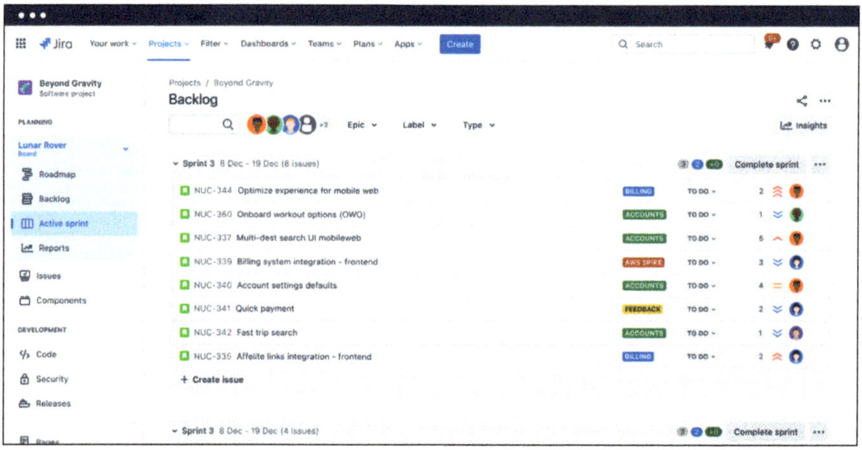

IMAGE COURTESY OF JIRA

FIGURE 8.6

Your final sprint plan should include the work, the estimate, and the person assigned to the work item.

Fine-Tune the Plan with Team Metrics

Tracking a few team-level metrics can improve the team's planning practice, increase your confidence in target dates, and help the team discover and overcome issues that slow down development. New teams typically require a few months to establish clear patterns, so it's important to begin tracking metrics early if you want reliable data for forecasting. The following metrics are popular ways to assess team productivity and improve planning.

- **Cycle time:** This tracks how long it takes the team to go from one workflow step to another, such as design to development or development to testing. Measuring this can help the team establish data-driven estimates using previous examples or help the team identify new ways to drive efficiency across steps.
- **Development velocity:** This reflects the typical amount of work a team can complete in a given timeframe and is usually measured in hours or story points. When the team's velocity isn't consistent or meeting expectations, planning go-live dates is difficult. If you notice inconsistent velocity or work that gets carried over across iterations, the team might be dealing with requirements issues, scope creep, estimation problems, unrealistic pressure on the team, or other culprits.

- **Defect frequency and root causes:** It's critical to track the number of defects, severity of defects, and reasons for defects found during preproduction testing and after production releases. Doing so will help you uncover and address root causes to improve quality earlier. It will also help the team set aside more time for testing or production support based on trends.

PRO TIP MAKE DELIVERY SKILLS A STRENGTH

If you are struggling to lead a team during product development, ask for help. (I did, and it was one of the best decisions I ever made.) You can usually find a few experts who thrive in this environment and can teach you a few more life-hacks to stay one step ahead. You can also explore resources from Agile Alliance, take a Certified Scrum Product Owner training, or simply start shadowing others whom you admire. The trick is to stick with it and turn execution into a strength because it will sharpen your knowledge, judgment, and communication skills and prepare you for future challenges ahead, big and small.

Prelaunch Validation and Testing

How can I validate that the solution meets expectations before launching?

After the development team codes the solution and feels confident that it meets business and user requirements, it's critical for the entire team to participate in the testing process. While testing strategies vary by company—including who tests, how and what they cover—it's the PM's job to prioritize quality and ensure that premarket testing happens before shipping a solution. There are many different forms of testing to accomplish this, but here are some of the techniques every PM should leverage.

Usability Testing

When you're launching a new product, major experience update, or new customer journey, usability testing is the best way to uncover blind spots and suboptimized parts of your product's design. You can run usability testing at any point, but running tests *prior* to launching something new gives the team a chance to identify and address problems proactively that will hamper customer success, negatively influence user perception, and ultimately degrade business metrics. For example, usability testing can reveal specific areas that cause customers to abandon your solution or begin switching to cost-intensive customer support channels.

To run a usability test, you'll need to create a set of predetermined tasks that you want to assess, build a thorough research plan, and recruit end users who are willing to test your product. Design partners can usually help with these tasks, but if you don't have internal support, explore usability testing platforms like UserTesting (as shown in Figure 8.7) or Maze for assistance.

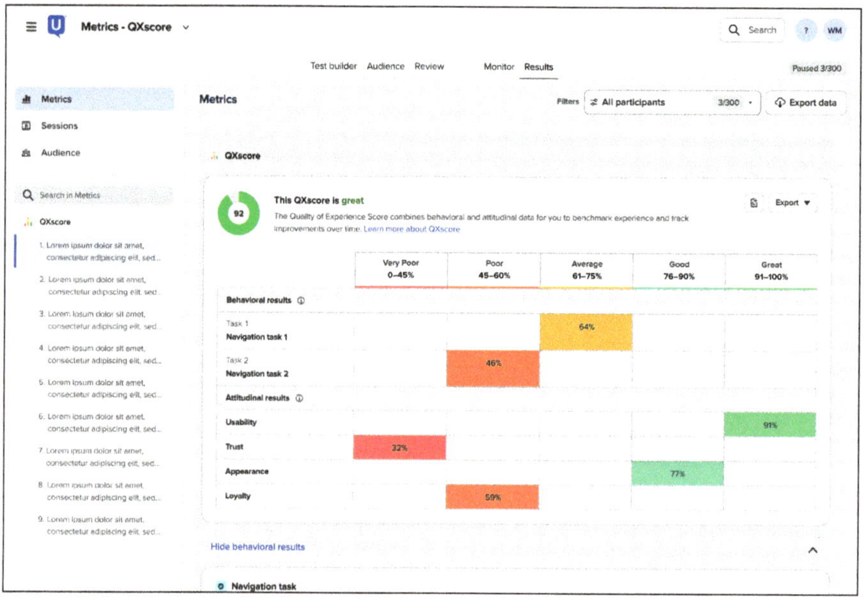

IMAGE COURTESY OF USERTESTING

FIGURE 8.7

User research platforms like UserTesting offer templates and features to accelerate the entire research process, from recruitment to synthesis.

During the tests, capture customer feedback and sentiment about the design, in addition to quantitative metrics like task completion rates, error rates, or the time it takes a user to complete a task. These measurements come in handy, especially when evaluating if a new design will outperform an existing design. If the tests reveal any issues that might severely affect success metrics, ensure that the team fixes them before launching. (You can add minor issues to the team backlog for a future iteration.)

Acceptance Testing

Once engineers code the solution, it's time for acceptance testing to assess product quality. Acceptance testing evaluates end-to-end functionality, data flows, and integrations in a production-like environment to help teams identify problematic defects caused by missed requirements or bugs. Teams can leverage automation tools, manual techniques, or blend both methods to accomplish acceptance testing.

On high-performing teams, acceptance testing is a team-wide effort, not just the responsibility of one person. To assist with testing, PMs should help the team form a test plan, participate in testing, and address and prioritize issues caught in testing. When functional and nonfunctional requirements are strong, creating a test plan is easy; you can pull directly from the user stories and documented scenarios to create test cases with pass/fail criteria. From there, the team can build scripts to create automated tests, but it's also important for you as a PM to run manual cases on your own. Doing so helps you spot defects and opportunities for future iteration.

As the team uncovers defects in testing, they'll want your input on prioritization or severity levels, so they know what's important to address now versus later. To prevent bottlenecks, you can create a clear structure or rubric (like the example in Figure 8.8) that makes it easy for anyone on the team to assign a priority level and understand turnaround times to resolve the defect. This will empower the team to open up defect tickets and immediately get to work on the most critical issues.

During testing, velocity will slow down when the team doesn't have access to mock data, resilient environments, or sufficient automation tools. When issues like these arise, collaborate with your engineering partner and leadership team to assess the impact with data and identify paths to overcome these hurdles.

DEFECT SEVERITY RUBRIC

Severity Level	SLAs	Sample Criteria
Critical	Must fix prior to release	• Blocks the user from completing a critical task. • Negatively affects conversion or retention rates. • Creates a major regulatory, legal risk, or brand risk to the company.
High	Must fix within 1 week after release	• Blocks the user from completing a critical task but workaround is available OR blocks the user from completing a noncritical task and no workaround is available.
Medium	Must fix within 30 days of release	• Blocks the user from completing a noncritical task but workaround is available.
Low	Fix when time and capabity is available	• Issue occurs infrequently or does not block functionality.

FIGURE 8.8

Tailor the defect severity scorecard based on your company's needs and risk appetite.

NOTE TESTING AI-POWERED EXPERIENCES

In Chapter 6, "How: Defining the Approach," you learned that teams building AI-based products leverage evals to test and assess the quality and performance of specific models. When your models integrate with other systems, services, or apps to power experiences, you'll still need to conduct acceptance testing to ensure that end-to-end functionality functions as expected. For example, if you build a customer support chatbot powered by an LLM, you should create evals to assess the quality and accuracy of model responses and then perform acceptance testing to verify that the presentation layer, orchestration services, data flows, and integration points work as expected.

Internal Product Demos

Demos aren't an official testing technique, but they're a great way to keep stakeholders aligned on the product's direction, gather feedback from important leaders, and celebrate progress or learnings. At a minimum, they can help you avoid surprises on launch day.

You can host a live demo or prerecord one and share out. In either format, the PM (or team representative) should start by setting context for the audience *before* showing the solution. Your intro should include the original problem statement, success criteria, expected benefits, and steps the team followed to design the solution. It's also important to explicitly call out goals that are *out of scope*. Once the audience has context, demo the product and call out key features, user benefits, critical integration points, and areas that involved significant partnership with dependent teams. You want the audience to understand how you balanced desirability, feasibility, and viability constraints. (FYI, glitches are a rite of passage for presenters during live demos, so do yourself a favor and prerecord a video to share in case a glitch happens. You'll thank yourself later!)

IMAGE COURTESY OF BOB RESELMAN

FIGURE 8.9
The demo gods will come for you—be forewarned!

After the demo, collect feedback from the group, but stay ruthless when prioritizing it to avoid scope creep. Use your best judgment to decide which feedback is essential to address now versus later and be transparent with stakeholders by sharing your rationale and decision before launch.

Pulling It All Together

Leading a team during development is a lesson in resilience and grit. Every hour, you're context-switching to provide clarity, enable velocity, test the solution, and keep the backlog healthy for upcoming iterations. And despite your very best efforts to keep the work on track and stakeholders informed, unplanned issues can emerge and threaten to throw the work off track. When this happens (and it will happen), don't take it personally. It happens to everyone! Stay positive, solution-oriented, and data-driven so you can make informed calls using all the customer and business insights available. And then examine it later to see if there's something the team can do better next time.

As hard as it is, leading a team through execution can be immensely rewarding. There's nothing better than working alongside a well-oiled, high-performing team marching toward a common goal that you believe in. And truth be told, launch day can feel like a great achievement for you and the team after being in the trenches for weeks. Speaking of launch day, let's cover a few extra steps in the next chapter to round out your delivery and go-to-market toolkit.

CHAPTER 9

Launching Products

Confidence levels fluctuate in the weeks before a product launch. Some days, the plan seems on track, the team is marching in the same direction, and you feel great about the upcoming release. Other days, you wake up in the middle of the night wondering why you didn't engage the marketing team sooner and if launch day will be a complete disaster.

The emotional rollercoaster is normal. It doesn't mean that you're a bad product manager; it just means there's a lot on your shoulders.

Truthfully, high-performing PMs don't know all the answers. They just know how to *get all the answers* by pulling in experts, being nosey, and pushing through ambiguity. Take product marketing, for example. Many PMs spend months thinking about product-market fit, projecting business outcomes, and crafting requirements, but know absolutely nothing about marketing strategies. That's totally fine! You just need to understand when to pull in marketing experts and empower them to do solid work. Or perhaps you've never worked on product pricing (also very normal). In that case, get nosey about pricing research methods, pull in the experts, and ask for help.

As a product builder, you're constantly facing new questions that feel impossible to answer on your own. This happens *a lot* during the prelaunch phase. Tenacity and collaboration can power you through most questions, so don't sweat it. Grab your partners and focus on answering the questions that matter most when planning for launch:

- What are customers willing to pay?
- What must be true to go to market?
- How can we ensure a smooth product release?
- What's everybody doing on launch day?

Go-to-Market Explained

World-class products (and critical features) don't generate thousands of customers overnight. You can have the best problem-solution fit in the world, but if customers can't find your solution, don't like the price, or can't understand your value proposition, you can kiss conversion goals goodbye. That's why every team needs a go-to-market strategy before launching a new product, entering a new market, or attempting a major pivot for an existing solution.

A go-to-market strategy is a plan that explains how your company will *attract* and *convince* customers to buy or adopt your solution.

Building a go-to-market strategy requires expertise across several domains, including product, technology, design, marketing, pricing, business analysis, branding, and distribution. Thankfully, there are experts who usually assist with this, a few popular growth and pricing strategies to consider, and a common way to evaluate the acquisition journey. Let's start with the experts first.

Go-to-Market Experts

Many product managers partner with marketing and growth experts to form a go-to-market strategy. That said, every company is unique and titles come and go, so if you don't see the exact roles below at your company, check with the marketing, technology, and product leaders to determine who typically drives these responsibilities and tasks for your company. And if you are the lucky person responsible for go-to-market planning, then read up on these roles and draw inspiration from them.

- **Product marketing managers:** These experts translate the product's value proposition into compelling messaging, a clear voice, and marketing strategies that generate awareness, build demand, and drive conversion. In many companies, they're responsible for defining and coordinating the final go-to-market plan. They often work hand-in-hand with product managers, technology teams, sales functions, and design experts to craft content, experiences, and campaigns that generate sales leads.
- **Martech experts:** Short for *marketing technology*, martech experts are the owners of capabilities that drive growth and manage the customer life cycle. They own customer-relationship management (CRM) solutions, customer data platforms (CDPs), analytics solutions, email marketing tools, and other automated capabilities to drive personalized communications with customers. They usually report to enterprise product teams, platform teams, or marketing groups. Every PM should have a deep understanding of the capabilities available to influence customer experiences, so find these experts and start building bridges with their team.
- **Pricing analysts:** These experts help companies maximize profitability and growth by deciding what to charge for products and services. They form pricing strategies by evaluating market trends, monitoring competitor products, and building financial models that forecast growth, expenses, and other factors that influence profitability. These roles usually sit under finance, business, or marketing groups.

- **Growth product managers:** This is a specialized product management role that focuses heavily on helping companies grow and scale efficiently. They tend to focus on optimizing customer journeys that heavily influence business metrics, such as the sales funnel or onboarding flow. They also identify new monetization or revenue strategies. In some cases, these roles may overlap with performance marketers, given the focus on acquisition, ROI, and experimentation.

Product-Led Growth vs. Sales-Led Growth

When you launch a new product, your first action is to assess product-market fit. Without it, growth won't happen. That said, it's still important to identify which growth strategy your company plans to pursue when forming a go-to-market plan, because this decision will impact budget, resources, and the customer journey. Today, many companies choose one of two popular strategies: sales-led growth or product-led growth.

- **Sales-led growth** is a model that relies on large sales teams spending significant time and energy to build personal relationships, demo solutions, and convince customers to buy a solution. This approach requires a hefty marketing budget to generate leads, build demand, and transform prospects into paying customers. Sales-led growth was a popular strategy during low interest-rate environments when companies prioritized customer growth over pretty much anything else and were comfortable having large marketing budgets.
- **Product-led growth** relies on the actual product itself to drive awareness, consideration, conversion, and retention. Instead of building demand through expensive marketing campaigns, product-led growth relies more on word-of-mouth and the actual product experience itself to gain market share. For example, companies that prioritize product-led growth strategies tend to offer free trials, freemium versions, an exceptional user experience, and self-service capabilities that deliver ongoing value for users to keep them engaged and become paying customers. In this model, users can explore the product directly. As companies tighten their belts and focus more on profitable growth (versus growth at all costs), more companies are investing in product-led growth.

Of course, both strategies can work depending on your market, industry, or solution. For example, product-led growth works well in B2C environments where a user can independently explore and receive value right away from a new product. Sales-led growth still works well in B2B environments that require complex integrations or contract negotiations. And yes, you can even have a hybrid approach to drive conversion across different segments.

Funnel Reporting

If you don't have a plan to measure customer acquisition, retention, and churn right away, you can't assess product-market fit. And if you can't assess product-market fit, you're not ready to launch. To get the team ready, you'll need to establish funnel reporting.

Funnel reporting is a technique used by marketing and product teams to distill, monitor, and influence the customer acquisition journey over the four common stages shown in Figure 9.1. Funnel reports help you identify inefficiencies or common drop-off points and track how well optimization efforts are improving top-of-funnel, middle-of-funnel, or bottom-of-funnel metrics. You can build them by using clickstream analytics tools like Google Analytics, Adobe Analytics, Mixpanel, etc.

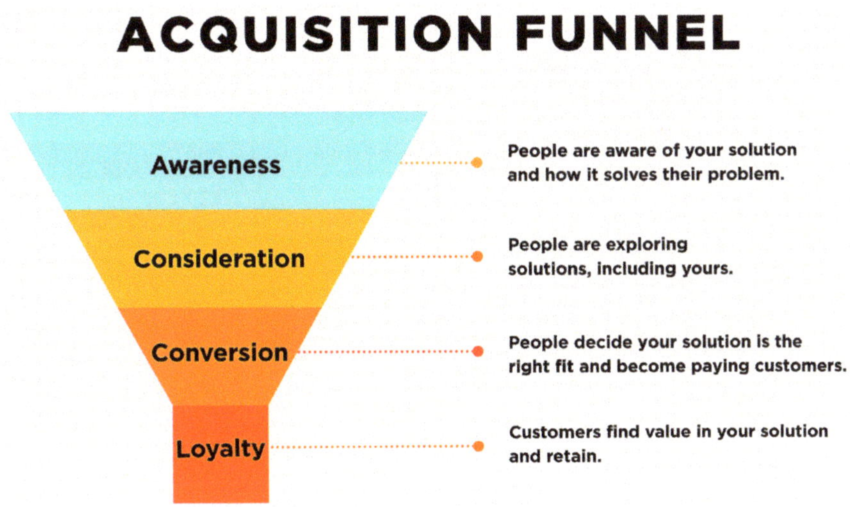

FIGURE 9.1
Your marketing funnel should represent the key stages and business metrics in your customer acquisition journey(s).

While funnels are useful, it's important to understand their limitations. Funnels often represent stages of the customer journey from the business's perspective. They look linear, straightforward, and easy-to-read during a business review. But customer journeys are usually nonlinear and more complex than you might expect. (Just show your business leaders a Sankey diagram if they don't believe you!) For example, users may visit your landing page, abandon right away, and come back weeks later after shopping around first. Or they might start the purchase process, get busy, and hit a session timeout. Or they might experience an error when adding something to the digital shopping cart. Or they might see a retargeting email from you, but instead of clicking the call-to-action, they navigate to a browser tab that's been open for days. The point is that your funnel represents *business* metrics, not your customer's journey. Don't confuse the two. When you want to understand why funnel metrics look a certain way, identify and evaluate common user journeys (using your own eyes) to learn what's actually happening from the *customer's* perspective.

And one more thing—*always* report on retention. It's easy for teams to focus on acquisition metrics during funnel reviews because business leaders love to talk about growth and conversion. I won't lie, it feels great to see conversion rates increasing! But if you're only looking at sales, you're forgetting the "fit" part of product-market fit. You want products that stick; retention and churn metrics are critical signals to ensure that growth is real.

PLAY 17

Selecting a Pricing Model

What are customers willing to pay?

When launching a new product or entering a new market, it's important to align on an attractive pricing model that drives consideration and conversion without undercutting or overselling your value. It's a delicate balance. You need to know what customers are willing to pay, and what your company is willing to charge. And while your pricing model will change over time as competition evolves and market trends shift, you shouldn't constantly A/B test your price for everyone to see in production. Doing so will condition your customers to sit around waiting for a better price. Instead of pricing tactics, you need a pricing strategy.

In some cases, the company staffs internal pricing teams to research and determine pricing strategies and pricing models for a product. This expertise is especially critical in companies that offer dynamic or model-driven pricing that assesses specific factors in real time to determine price. In other cases, the company might hire a consulting firm for advanced pricing research, testing, and analysis. But if you happen to find yourself in the driver's seat and you need to lead the charge on pricing strategy, here's how to get started.

Use When

You need to define a go-to-market pricing strategy for a new product.

Average Time

At least 2 weeks for initial analysis and customer research

Try This

1. **Assess your business and competition first.**

   To determine your pricing model, start by collecting data on product costs, goals, and what your competitors are doing. Compile the following:

   - The actual costs to create your product, including the direct costs to build and ongoing costs to maintain it. (If you did the technical feasibility assessment in Chapter 6, "How: Defining the Approach," grab costs from this exercise.)
   - Business goals, such as your revenue targets or profitability targets. (If you projected impacts and success metrics from Play 11 in Chapter 7, "Establishing a Plan and Getting Buy-in," grab the goals and targets.)
   - What competitors are charging for similar solutions. (If you completed Play 3 in Chapter 3, "Why: Defining the Problem and Opportunity," pull up the pricing info. Otherwise, gather intel on what competitors are charging.)

2. **Collect customer feedback on price points and pricing models.**

   Once you know the business goals and competitive landscape, you'll need to assess how customers perceive the product's value, the product's value *compared to competitors*, and whether price elasticity exists. (Price elasticity is just a fancy phrase to assess how sensitive consumers are to price changes in your product.) Combined, these inputs help you pinpoint an acceptable price range for customers and a pricing model. Figure 9.2 shows popular research methods you can use to gather this data.

PRICING RESEARCH TECHNIQUES

RESEARCH METHOD	TYPE	DESCRIPTION
Customer Interviews	Qual	Gather market feedback to compare pricing and feature value against competitor solutions.
Conjoint Studies	Quant	Simulate trade-offs to determine value of specific features or product configurations.
Gabor-Granger Method	Quant	Determine optimal price point and price elasticity by assessing willingness to pay at different price points.
Van Westendorp Price Sensitivity Meter	Quant	Determine acceptable price range and optimal price point with a four-question survey.

FIGURE 9.2
Product teams should start pricing tests early, ideally once you have a clear value proposition and feature set.

If you're not sure where to begin, start with customer interviews to collect feedback on your product's price points. Research how consumers perceive your product's benefits compared to competitor solutions. Many teams run a four-question survey using the Van Westendorp Price Sensitivity Meter method to determine an acceptable price range (as shown in Figure 9.3).

While you're assessing price point ranges with consumers, go one extra step and gather user feedback on pricing models as well. Unless you've been living under a rock, you probably know that pricing models vary a ton. Figure 9.4 shows common pricing methods by industries.

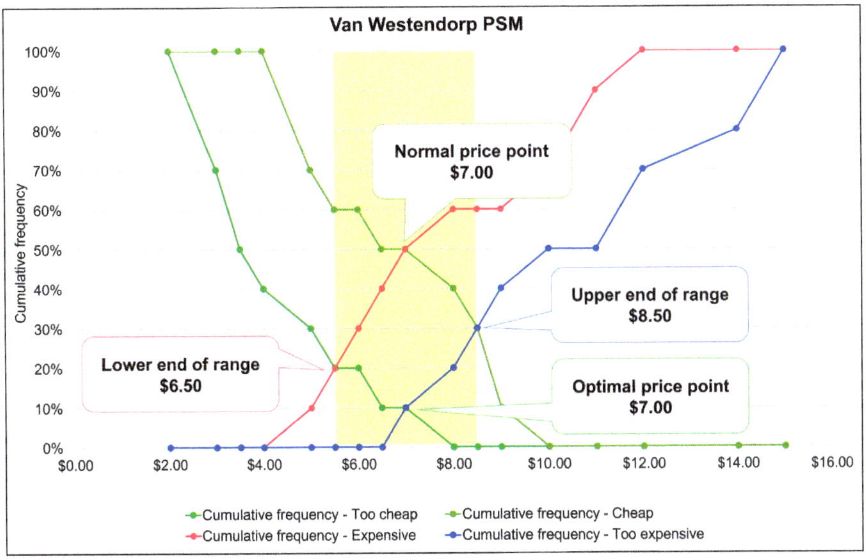

IMAGE COURTESY OF CONJOINTLY

FIGURE 9.3
Tools like Conjointly, Sawtooth, or SurveyMonkey offer advanced surveying and analytical capabilities to inform pricing decisions.

COMMON PRICING METHODS

TYPE	HOW PRICING WORKS	COMMON USES
Cost-plus	Based on production costs, plus a fixed percentage for profit	Manufactured products
Value-based	Based on what customers are willing to pay	Luxury goods
Outcome-based	Based on measurable outputs or success metrics that matter to the customer	AI agent solutions
Algorithmic	Based on trends in supply, demand, or other unique market factors	Airlines, hotels, car insurance
Usage-based	Based on actual consumption of product	Cloud services, utilities
Tiered Structure	Based on different product configurations, such as basic, standard, or premium packages. May include freemium offerings.	SaaS products with retail & enterprise customers
Pay-per-seat	Based on number of units, licenses, or users with access	Enterprise workflow tools
Subscription-based	Based on payment across regular intervals (i.e., monthly or annually)	Content providers

FIGURE 9.4
Many companies explore hybrid options across one or more of these models.

3. **Determine your go-forward pricing and promotional strategy.**

   After research, compile pricing feedback, competitor data, and internal business analysis. Then review the key insights with business leaders and finance partners. Start by sharing the data on what customers are willing to pay. Then share what the company's goals are and how market conditions (such as competitor pricing) could influence customer behavior. After the team reviews the data, finalize the following:

   - The MVP pricing strategy, including the price points and pricing model by segment
   - A plan to monitor the pricing strategy post-launch, including timelines and metrics to collect
   - Any promotional pricing strategies you'll use to attract and convert your initial customers, such as discounts, bundles, or flexible payment options
   - How the team can adjust pricing changes post-launch using experimentation or configuration services

4. **Integrate pricing requirements into the product's functionality.**

   Update the product requirements and data strategy to ensure that your MVP can support the desired pricing and product configurations. For example, if you're offering a pay-per-seat model, then ensure that the product restricts new "seats" over a specified limit.

LEARN MORE

PRICING TECHNIQUES

Some people spend their entire careers mastering pricing research and strategy. Lean on company experts when you can, but if you want to learn more, start with these resources:

- ***Monetizing Innovation*** **by Madhavan Ramanujam and Georg Tacke**
- ***Scaling Innovation*** **by Madhavan Ramanujam and Eddie Hartman**
- ***The Strategy and Tactics of Pricing*** **by Thomas Nagle, John Hogan, and Joseph Zale**
- **Case studies and pricing strategy articles from Qualtrics, SurveyMonkey, Sawtooth Software or Conjointly**

PLAY 18

Building a Go-to-Market Plan

What must be true to "go to market"?

Now that we've covered product marketing, growth, and pricing concepts, you're probably wondering what a go-to-market (GTM) plan actually includes? Truthfully, it's different everywhere you go. Typically, a GTM strategy articulates how customers will find you, what messages you'll lead with, and what the details are about pricing strategy. Some companies also include operational components, especially when there are cross-channel customer touchpoints across digital, voice, or in-person interactions. With so many factors to consider, it's common to include product, marketing, martech, pricing, operational, and finance experts in the GTM planning process.

Given the extreme nature of coordination, senior leaders often drive go-to-market planning and decisions. That said, you might find yourself leading and assisting with go-to-market planning for a new product, a major pivot to an existing product, or even an internal product rollout. If you do, here's how to demonstrate excellence when aligning and communicating across different functions.

Use When

You're introducing a new product into the market, pivoting an existing product, or rolling out an internal solution. (You don't need this for individual feature launches.)

Average Time

2 weeks to make decisions, plus time to execute

Try This

1. **Determine product positioning, messaging, and branding with marketing experts (or hire an expert).**

   If you're lucky enough to have a large budget and professional marketing team, they'll lead this part. But if you're on a scrappy team, find ways to accomplish positioning, messaging, and branding to give your product a shot at success. Here's what each one includes:

- **Product positioning** is an exercise to understand and influence how a market thinks and feels about your product. Craft a positioning statement that guides your marketing activities. The statement should answer who your unique market segments are; the unique traits, features, or benefits you want to emphasize to each segment; and why you're better than other solutions.
- **Product messaging** is the actual story or message you show prospects. Write it using plain speak (not technical jargon). It should explain why prospects should care about the product, what it accomplishes for them, and why they should believe you. Tailor it to different segments based on what matters to them.
- **Product Branding:** If you're launching something entirely new, select a logo, name, and brand voice to form a unique identity. Consider branding "hero features" or trademarking capabilities that differentiate yourself from competitors.

2. **Define your customer acquisition strategy.**

   Acquisition strategies are complex and usually mix tactics, channels, messages, and capabilities to attract and convert customers. Your acquisition strategy should answer these questions:

   - What is the overall acquisition strategy to drive qualified leads and prospects into the funnel?
   - What are all the techniques and tactics we'll use across each stage of the acquisition journey? And how does it vary by segment?
   - Who is ultimately responsible for managing the acquisition plan and activities?
   - What is the official growth strategy? (For example, sales-led, product-led, something else)
   - How much budget is available to invest in acquisition efforts?

   If you're launching an MVP, cherry-pick a few tactics from Table 9.1 until you validate product-market fit.

3. **Finalize the GTM plan and track progress.**

   Figure 9.5 shows all the elements required to form a complete GTM plan. Your final GTM plan should articulate a decision for each item. In reality, this can take weeks to finalize, so keep it updated as you complete follow-ups. Anyone who reads it must easily understand how the company plans to attract prospects and convince them to buy.

TABLE 9.1 ACQUISITION FUNNEL DECISION FRAMEWORK

Stage	Decision Needed	Techniques to Consider
Awareness	How will customers discover your product?	• Omni-channel marketing campaigns • Search engine, answer engine, and generative engine optimization (SEO, AEO, and GEO) • Cross-sell offers • Retargeting journeys
Consideration	How will customers understand your product?	• Product content • Industry white papers • Free trials • Demos
Conversion	How and where will customers purchase or convert to your product?	• Pricing tests • Promotional offers • Sales funnel optimization • Flexible payment methods
Loyalty	What will customers experience with the product and how will they engage over time?	• Activation journeys • Engagement-based features • Customer support touchpoints • Renewal and loyalty offers

FIGURE 9.5
Prelaunch status updates should include product development updates and GTM readiness updates.

PLAY 19

De-risking a Release

How can we ensure a smooth release?

While you're busy clearing acquisition, pricing, and GTM hurdles, it's still important to keep development on track and ensure that the team de-risks potential issues before launch. Of course, the worst thing that can happen is discovering that your product doesn't solve an actual need and therefore your product-market fit isn't possible. The second-worst thing that can happen is that the launch is so bad you can't get legitimate product feedback, and the team starts spiraling.

Trust me, unplanned issues will pop up after launch. Part of the PM's job is to preempt and mitigate them. I realize some company cultures (usually startups) say, "move fast and break things." But if you're supporting a midsized or large business, the company has more at stake: existing customers to retain, a brand to preserve, and perhaps more regulators to manage. All that means more to protect. So, whether you like it or not, risk management is part of product management. That doesn't mean risk management has to be heavy-handed; it just needs to be thoughtful. Here's how to strike the right balance.

Use When

You're launching a brand-new product and need to minimize risk.

Average Time

2 hours to plan, plus time to implement recommendations

Try This

1. **Build a phased roll-out plan with success criteria.**

   One of the easiest ways to manage risk is by phasing product rollouts. Before a product launch, consider controlling the initial user population and agreeing on specific outcomes that must be true before the team can expand, such as low error rates. Here are four ways to control your launch population and quickly mitigate kinks before a bigger release.

   - **Conduct an alpha test.** Provide coworkers with access to the product and collect feedback before releasing to real customers. Alpha testing helps the team uncover and resolve bugs and usability issues before a public release without risking the brand or lost sales.

- **Launch a beta test.** Find real-life users to gather feedback before a broader release to the public. Beta tests are especially helpful because they occur in production and can uncover networking issues or platform reliability issues that you won't see in internal testing. To run one, determine how many users you want to include, provide a sign-up method, and establish a way for them to submit feedback and bugs.
- **Stage rollouts.** When launching a new product or big enhancement to *existing* users, throttle or control how many people experience the solution for a few weeks. For example, expose 10% of existing customers to a new solution for two weeks and monitor usage and platform metrics to detect issues. Use feature flags to accomplish this.
- **Launch an experiment.** A/B testing might be appropriate when launching a new experience change or feature, but it's not right for *everything*. (We'll cover a whole method on that in the Chapter 10, "Managing Products In-Market.")

2. **Set up feedback loops.**

   Before launching a product, establish clear feedback loops for users and stakeholders. Feedback loops can be manual or automated, but clearly document and socialize them, so the team gets actionable feedback right away.

   - **Establish a bug reporting process.** Do not launch without a bug reporting process…ever! A solid bug reporting process includes a clear way for customers or stakeholders to submit feedback, a process for triaging bugs, service-level agreements (SLAs) that dictate resolution guidelines, and a dashboard that monitors bug status and trends. If you use customer support teams, partner to streamline feedback and capture volume for each issue.
   - **Integrate session replay and clickstream analytics tools.** Session replay tools (like Fullstory, Quantum Metric, etc.) are a must-have for product teams, and every PM should know how to use them. They show you what's happening on the user's screen, and many offer AI-powered summaries to triage defects and errors quickly. Clickstream analytics tools allow you to build funnels or track specific events in the user journey at an aggregate level. It's important to monitor clickstream

reporting and watch session replays daily because most users will abandon your product rather than report a bug. Consider building a rotation schedule so that everyone on the team learns these tools and builds empathy for customers.

- **Make dashboards public.** The second you launch, expect loads of questions and little time to answer them. Do your future-self a favor and set up reporting dashboards that anyone can access. (Finding ways to scale information without being the bottleneck gives you more time to focus on the work.)
- **Be present.** If you're a PM launching an internal solution or workflow tools, be physically present on launch day. Being present builds trust and allows you to notice new details that can affect their overall experience, like their monitor size or workplace environment.

3. **Give your team space to fix things.**

   Despite your best intentions, things happen. (It's normal!) The trick is to avoid scaling problems. This means instead of promising internal stakeholders a tranche of new features right after launch, proactively set aside capacity for the team to fix issues and keep tech debt to a minimum. For example, allocate 50% of the team's total capacity for production support for three weeks after a major launch. If you don't need it, great! But if you do need it, this protects the team from an unfair capacity squeeze.

Launch Day

What's everyone doing on launch day?

Launch day can feel exciting. It can also feel a bit... anti-climactic. If you built a solid go-to-market plan and followed the de-risking methods in this chapter, the first few hours might be somewhat quiet as you wait for data to roll in. In fact, here's what a typical launch day might look like:

- 6:30 a.m. The cross-functional team meets in a physical room for the day or begins an all-day virtual meeting. Ideally, someone brings donuts.

- 7:30 a.m. Engineering confirms the production release happened. The team members perform smoke testing in production to ensure that the product works as designed.
- 7:55 a.m. The product and tech lead "sign-off" on the release and note the time. The team celebrates with fire and hand-clap emojis.
- 8:30 a.m. The PM sends a brief release announcement to a small group of immediate stakeholders. (The larger announcement doesn't go out until a few days later, just in case the team must roll back the release.)
- 9:00 a.m.–4:00 p.m. The team spends the rest of the day together monitoring observability tools and waiting for the first users to check out the product. Pizza is provided.
- 4:45 p.m. Engineering checks the logs. Someone bought the product! The team celebrates with party parrot emojis and GIFs.
- 5:01 p.m. Customer support team reports the first bug. The customer never received a confirmation email of their purchase.
- 5:30 p.m. The team identifies the root cause. A feature flag wasn't turned on. The team immediately turns it on and fixes the issue in production.

For proactive teams, launch day usually isn't too bad. The subsequent days grow more hectic as volume and customer feedback increase. But savvy PMs don't sweat it because they've operationalized feedback loops, established a bug process, and set aside team capacity to handle it. For teams that take quality shortcuts and bypass thoughtful planning, launch days aren't so easy. If you're on a team that experiences painful launch days, consider pulling the leads together and hosting a retro to find ways that make release day more peaceful.

Pulling It All Together

Product launches directly reflect your partnership skills. I've never met a product manager who launched a solution for their company entirely on their own. But I've met dozens of PMs who launch products by collaborating with partners all the time. The best product managers master product releases by building operating models that build clarity and confidence for everyone involved.

That doesn't mean you must be everywhere, doing everything. In fact, you won't be able to scale yourself. Think back to earlier concepts in this book. Your product vision should inform marketing's work. Your metrics plan should inform the funnel creation. Your bug severity scorecard should help the team assess and prioritize defects quickly. The ways of working audit should clarify how the team functions. See the pattern? All these artifacts, concepts, and activities build on themselves and prepare you for the biggest challenge yet: managing an in-market product.

CHAPTER 10

Managing Products In-Market

In 2012, Stewart Butterfield stood before 50 employees and announced they were closing shop on *Glitch*. Their multiplayer online game had amassed tens of thousands of users, but the numbers weren't adding up. A core group of avid players loved the game, but new players weren't sticking around. Backed by $17 million in funding, the company had spent 18 months experimenting with new features and strategies, and it hadn't been enough.

With a few million dollars left in funding, the team started brainstorming new business ideas. Someone proposed spinning off the company's homegrown internal chat system as a commercial product. It was scalable, it helped remote workers connect, and it kept a transparent record of team decisions, which made onboarding teammates easy. If it worked for their company, surely other businesses would find it valuable, too. Within a week, they went all-in on the pivot.

That internal chat tool ultimately became Slack, the workplace collaboration tool with millions of daily users when Salesforce acquired it for over $27 billion.

Slack's story is a lesson in *iteration*. After launch, your product becomes a living organism that needs constant attention and nourishment to survive in a world where customer preferences evolve and competitive threats emerge. In some cases, you can pour everything into the concept and still fail. In other cases, you'll succeed after dozens of iterations and pivots.

The trick to iterating successfully is using the most precious gift you have—in-market feedback. Of course, every product manager needs a compelling long-term vision, strategy, and roadmap, but if you ignore what's happening in-market, you'll scale problems, limit your potential, and miss an opportunity to drive meaningful change. So, if you've just launched a new product, feature, or solution, start with data and let it guide you across the questions that matter most:

- Do I have product-market fit?
- How is our product performing?
- How can I prepare for growth and scale?

Finding Product-Market Fit

Do I have product-market fit?

In Chapter 2, "The Effective Product Manager's Toolkit," we defined product-market fit (PMF) as an *outcome* that occurs when a specific market adopts your solution *and* the solution has a clear path toward sustainable business value. Accomplishing PMF should be the first goal for every team after launching a product.

If you're hoping for a playbook method on how to achieve PMF quickly, you won't find it here. Attaining PMF is not a straightforward process. Instead, entrepreneurs and product builders will tell you it's a long journey that demands countless cycles of iteration. In fact, when product expert Lenny Rachitsky interviewed 24 popular B2B startups, most companies didn't report having product-market fit until *two years* after launching their first solution. Some took even longer; Miro and Figma reported taking nearly five years to accomplish product-market fit.[1] The journey to PMF is an exercise in tenacity and patience all at once.

0-to-1 Products

When launching a new product, you're entering the 0–1 phase. This stage represents going from initial concept to minimum viable product. In this phase, you must fine-tune the solution until seeing clear evidence that a *specific market* values the product. And having a single user or paying customer doesn't count—it needs to be a *market* you can easily identify with a set of common attributes.[2]

During this phase, obsessively watch customers use the product, collect feedback, and improve the product. Learn which hypotheses were totally wrong, form new ones, and look for signals you're getting close to PMF (see Table 10.1). It's important to bring the team together weekly to examine in-market data, challenge core assumptions, and decide what to test. Hypothesis-driven roadmaps should be your best friend.

---

1 Lenny Rachitsky, "Finding Product-Market Fit," *Lenny's Newsletter*, October 27, 2021, www.lennysnewsletter.com/p/finding-product-market-fit

2 Common attributes might include demographic, psychographic, behavioral, and geographic data points. Aim to establish segments that share needs and preferences so you can develop effective marketing and product communication strategies.

TABLE 10.1 PRODUCT-MARKET FIT SIGNALS

Signals You *Might* Have PMF	Signals You *Might Not* Have PMF
• You retain paying customers. • Users would be disappointed if your product went away. • Sales and marketing efforts are converting customers effectively. • Your user base and revenue are growing steadily. The growth chart starts to look like a hockey stick. • Your product is starting to grow organically (by word of mouth). • You're struggling to keep up with demand.	• Retention is poor, and you see customers cancelling, refunding, or not renewing at high rates. • Prospects sign up for freemium trials, but don't convert at the end of the trial. • Your customer acquisition costs are high, but your customer lifetime value appears low. • Your user base is small and growth is stagnant. • Customer feedback underperforms industry benchmarks.

1-to-n Products

After finding PMF, companies start looking for ways to expand the product's reach. Product teams call this phase "1-to-n" product management because the strategy shifts from finding the minimum viable product to scaling the solution. At this point, product managers typically focus on driving growth and preparing for scale. Roadmaps often include strategies to optimize conversion, onboarding, retention, scalability, speed to market, and product quality.

During 1-to-n product management, it's normal to become laser focused on specific improvements and operational efficiency, but don't get so entrenched in tactical items that you forget the bigger picture. While you're enhancing and optimizing the product, new technologies emerge, customer preferences shift, and your competitors try to outpace you. If you don't keep a pulse on overall PMF and viability, you'll miss the forest for the trees. You can't optimize your way out of a busted strategy, so practice these behaviors to keep PMF and viability in focus.

1. **Tie every product enhancement to a customer and financial metric.** Every feature, platform change, or experiment should somehow strengthen PMF or business viability. For example, if you're considering an enhancement to increase feature adoption, assess how that impacts customer acquisition, retention, revenue, or costs.

2. **Establish consistent forums to review product viability and performance data.** If you're hosting product reviews, don't just focus on new features, bugs, and recent releases for your specific journey. Include a slide that discusses market share, monthly recurring revenue, customer acquisition cost, customer lifetime value, and retention rates.
3. **Discuss the competition.** As a team, it's critical to review industry satisfaction benchmarks, discuss differentiation techniques, and assess how your product is performing relative to existing and emerging alternatives.
4. **Have the courage to pivot.** Many companies find great success after making a strategic pivot (see Figure 10.1). When a major trigger or insight threatens your product's future, revisit your strategy and pivot if needed.

Company	Started as a...	Succeeded as a...	Because they pivoted the...
Play-Doh	Wallpaper cleaner	Popular children's toy	Target user
NETFLIX	DVD mailing business	Content streaming service	Business model and technology
YouTube	Online dating site	Universal video-sharing site	Target user
amazon	Online bookseller	E-commerce marketplace	Value proposition
twitch	General live-streaming platform	Gaming-specific live-streaming platform	Target user

FIGURE 10.1
Product iteration and adaptation are survival skills in a competitive world.

EXPERT ADVICE WITH DAN OLSEN

ITERATING TOWARD PRODUCT-MARKET FIT

Dan Olsen is a world-renowned product management consultant, trainer, and author of The Lean Product Playbook. *He also runs a 13,000-person product meetup. After seeing hundreds of products succeed and fail, here's the must-know advice he has for teams seeking product-market fit.*

More than 80% of new products and features fail because they don't create enough customer value. And they don't create enough customer value because teams don't conduct enough research and validation work before and after launch to fine-tune the product and truly accomplish product-market fit. Of course, teams are often limited to qualitative techniques during prelaunch stages, but once you launch a product, it's crucial to continue research and validation efforts. Most products don't have instant product-market fit, so baking in time for research and iteration after launch is imperative.

After you launch a product, study quantitative data to understand who is signing up, where users drop off, what they click on, what they use, and what they don't use. The top indicators of product-market fit are repeat usage and strong customer retention rates. But don't just rely on aggregated quantitative data to assess performance. Dig deeper by pursuing qualitative user research and cohort analysis to understand your market. Find the avid users and understand what resonates most with them. Find the users who abandoned

Monitoring Product Performance

How is our product performing?

Your product's performance reflects hundreds of decisions you made when defining, building, shipping, and managing the product. Of course, it's unrealistic to track every single decision with data, but you can't build in-market roadmaps based on gut instincts, either. Instead, every PM needs a rinse-and-repeat way to holistically monitor product performance.

To make it simple, establish a routine process where the team evaluates performance across three buckets: customer success, business success, and product quality. Each bucket directly connects to the desirability, viability, and feasibility pillars covered in Chapter 2, and will help you form a balanced perspective when assessing product performance.

your product and learn why your product fell short of their expectations. Look for insights in the qual and quant data and iterate until you have true product-market fit.

I'll give you an example. Intuit built Quicken as a personal finance tool to help everyday consumers balance checkbooks and pay bills. But through user research, Intuit discovered that an increasing percentage of customers used Quicken to manage their small business finances. This was a major epiphany. It helped Intuit expand the market for Quicken and ultimately fueled a multimillion-dollar revenue opportunity. Without talking to users, Intuit would have missed this insight completely.

My other advice is this—don't start focusing on growth until you've validated that you have achieved product-market fit. I see this mistake all the time. If the data doesn't indicate strong usage rates, customer retention, or positive customer feedback, you probably need to iterate or pivot your product. Pivoting might mean you need to pursue a different target customer or change the core value proposition. Or add or kill features. Or change something about the design. If you find yourself in this situation, study what the research and in-market data is telling you and identify the core hypotheses you want to change. Once you determine a new direction, start the process again of validating the new hypotheses using qual and quant research techniques. And don't be afraid to pursue a new direction—it's common to pivot a product after initial launch.

Assessing Customer Success

You can conduct the best user research in the world, but your product or feature is still a hypothesis when you launch it. The only way to know if your solution truly solved a problem for customers is by collecting and reviewing data that reflects *their* definition of success. Ideally, your original problem statement defined customer success up-front, so you know exactly what to measure post-launch. But if you're shipping multiple changes at the same time or managing the entire product experience, it's best to evaluate customer success holistically. To do this, collect and evaluate data across utility, usability, and desirability, as shown in Table 10.2.

If you're managing a product with many customers, consider analyzing these categories *in aggregate and by cohort* to understand how different customer segments perform. Supplement the data with context by reviewing session replays, reading customer complaints, and running through the product experience yourself, including happy and unhappy paths.

The output of your assessment should include a list of data-backed pain points and opportunities that become roadmap candidates for future iterations. Categorize the insights into product-market fit issues, functionality gaps, user experience issues, or customer satisfaction drivers. Then size each of them and start the product development process over again by drafting a new problem statement to guide solution discussions. After you implement a change or improvement, reassess the data to study impact.

TABLE 10.2 CUSTOMER SUCCESS METRICS

Theme	Key Question	What to Measure
Utility	Do customers find your solution useful and valuable?	• Product/feature adoption rates • Product/feature engagement and usage rates • Importance-satisfaction scores by feature • Time to value
Usability	Can users easily accomplish their goals using your solution?	• Task success rates across journeys • Time it takes to complete task • Error rates • Customer support contact rates
Desirability	Do users like your solution?	• Customer satisfaction scores (NPS, CSAT) • Customer sentiment, ratings, or reviews • Customer complaints • Customer loyalty/average tenure

Assessing Business Success

Tying your work to business outcomes is an essential skill in product management. Doing this early and often also prepares you for product leadership roles that expect you to track, understand, and influence business performance across the entire customer journey. Table 10.3 shows the most common business success metrics you can use to measure product performance. Each industry varies, however, so tweak this based on what's right for your company and industry.

While each metric in Table 10.3 is important, it's unlikely that you'll care about all of them equally. (If you do, you probably have a clear strategy for your product.) Think back to the original business goals that shaped your problem statement or strategic opportunity—for example, does the company want to achieve growth in a new market, deepen share of wallet for an existing market, improve operational efficiency, or something else? Whatever the goal is, focus on the business metrics that *best* represent those original goals, so you can compare current performance to targets and adjust accordingly.

TABLE 10.3 BUSINESS SUCCESS METRICS

Theme	Key Question	What to Measure
Acquisition	How efficient are we at acquiring new customers?	• Funnel efficiency and drop-off rates • Conversion rate overall • Conversion rate by product configuration or pricing model • Conversion rate by channel • Customer acquisition cost (CAC)
Engagement and Retention	Are we retaining and growing our customer base?	• Retention rate • Churn rate • Customer growth • Active users
Revenue and Profitability	Is our product financially healthy?	• Monthly recurring revenue (MRR) • Average revenue per customer (ARPU) • Total cost of ownership • Gross and net profit margin • Customer lifetime value (LTV)
Industry Standing	How are we doing compared to the competition?	• Market share • Industry benchmarks on NPS/CSAT • Brand perception scores

Assessing Product Quality

Companies pay for quality in two ways: they either invest in quality up-front during development, or they pay for it later when bad quality causes lost sales and churn. Of course, you shouldn't fixate so much on quality that you reach diminishing returns, but you can't ignore quality's impact on business performance either.

You can strike a good balance by including quality metrics during product reviews, alongside customer and business metrics. Table 10.4 includes common metrics you should use to assess product and system quality throughout the product life cycle. If your company culture underinvests in quality, extrapolate these metrics into dollar outcomes and report on those, too. For example, what is the defect rate costing you in engineering salary dollars? How many lost sales did you incur from outages? Is latency causing high rates of funnel abandonment?

TABLE 10.4 PRODUCT QUALITY METRICS

Theme	Key Question	What to Measure
Software Quality	How often do we generate defects or errors?	• Model recall and accuracy • Model drift • Production defect rate • System failure rates • Test coverage
Recoverability	How well can we handle system failures?	• System availability • Mean time to detect an incident (MTTD) • Mean time to repair (MTTR)
Performance and Scalability	Can our product and systems handle an increase in demand?	• Response times • Latency • Throughput (TPM [transactions per minute] TPS [transactions per second]) • Cost per transaction

When quality issues cause harm to customers or business goals, quantify the impact, adjust the roadmap accordingly, and begin dedicating resources to fix issues immediately. Ignoring them only leads to more tech debt later. And be prepared to confront and resolve systemic root causes that lead to poor quality. It's tempting to assume quality issues arise from feasibility and technology choices, but common culprits include speed-obsessed cultures, limited understanding of customer behavior, and incomplete requirements. (I wasn't harping on nonfunctional requirements in Chapter 5, "What: Defining Solutions," just for fun!)

SURVIVING DATA ISSUES

Companies aren't perfect, data is messy, and observability tools can be challenging to use. But guess what? None of those issues stop your boss from asking you to send business impact analysis roughly two seconds after your release.

Yes, in an ideal scenario, you have the best tools, amazing data partners, and a perfect reporting rhythm. But many PMs do not operate in ideal scenarios. They operate in the *messy middle* of companies that pretend to be data-driven but don't have the full maturity to *be* data-driven. If you're in one of these companies, here's how to respond to common scenarios:

Scenario 1: Your team doesn't consistently review performance data.

If you don't have routine business reviews or product performance updates with critical stakeholders, establish a biweekly check-in where you present customer success, business success, and product quality metrics to partners. Make the meeting transparent, public, and factual. Focus on the most important metrics first. Assign follow-ups or host separate syncs when the team needs to swarm on a critical issue.

Scenario 2: Your company is a data desert.

When your company lacks observability tools, reporting product performance is nearly impossible. Immediately add robust analytics requirements to the backlog and bump them right to the top. Until then, align on 3–5 metrics that matter most and find proxies or use assumptions with caveats until you start producing and collecting data.

Scenario 3: The numbers are disappointing.

When metrics head in the wrong direction, be transparent and ask for input. Don't shy away from sharing data that will disappoint others or stump the team. Customer behavior is confusing and unexpected sometimes. Many heads are better than one.

Scenario 4: The data looks wrong.

If the data seems downright wrong or feels off, ask engineering and data partners to validate data that flows are working properly across systems. (Data mapping and sequence diagrams are especially useful when this happens.) You can also ask the team to automate test cases so that everyone knows when something breaks.

Driving Growth and Adoption

How can I prepare for growth?

I've never met a business leader who didn't want growth. But growth before product-market fit can be an expensive mistake that sends your customer acquisition costs soaring without financial benefits to offset the expense. To know if you're ready for growth, ask yourself three simple questions:

- Have we observed product-market fit?
- Do we have a path toward sustainable business value?
- Are we still committed to the original business goals, target markets, and strategic beliefs that shaped this opportunity?

If the answer is yes to all three, it's safe to pursue growth. The first step is simple: start by defining what growth truly means to the company. Some leaders want to see top-line revenue growth, some would rather focus on profitable growth, and others might simply mean they want more customers. Each of these comes with its own set of strategies and tactics, but for now let's focus on customer growth.

Retaining Existing Customers

In many cases, the best growth strategy is to focus on improving the product first. Product-led growth is cost-efficient, scalable, and ensures that your value proposition aligns to what customers *actually* experience. It also retains *existing* customers who generate a gold mine of data you can use to improve the product. Plus, you can't grow market share if folks are leaving you in the dust, so consider enhancing your core product and boosting retention before investing in massive paid growth. Here are a few ideas to consider.

- **Improve the core product experience:**
    - Eliminate usability or quality issues that generate customer support contacts.
    - Automate high-friction or frequent tasks to reduce cognitive load.
    - Establish embedded distribution tactics or growth loops that expand the value proposition and drive compounding word-of-mouth growth (i.e., user-generated content, shareable content, referral rewards, ratings and reviews, etc.)

    - Introduce benefits that increase loyalty (i.e., special access, recognition experiences, etc.).
    - Build an ecosystem of complementary, interoperable solutions to increase stickiness.
    - Leverage customer data to create relevant, personalized product experiences across onboarding, retargeting, and cross-sell journeys (i.e., next best action models, recommendation engines, etc.).
- **Use feedback from current customers to evolve the offering:**
    - Launch in-market experiments and beta tests with existing users to test new feature concepts.
    - Research why existing customers chose you and what they love about your product. Learn why some disengaged or canceled. Use this data to build customer profiles and fine-tune the value proposition, pricing, and positioning strategy for future GTM planning.
    - Explore customers' adjacent and aspirational needs. Use the insights to inform new product concepts or product expansion strategies.

Acquiring New Customers

Ideally, your product delivers strong organic growth before making huge investments in paid growth. (Remember, the product itself should be your core growth engine first.) That said, when you're ready to take the plunge, there are loads of strategies, tactics, and methods to attract new customers. In most cases, marketing professionals and growth specialists will guide this work. Here are a few strategies they'll consider with you:

- **Assign a team to support conversion rate optimization (CRO).** Optimize the sales funnel with a team that focuses on rapid experimentation.
- **Assign a team to support onboarding and activation.** Optimize onboarding journeys to reduce time to first value (the first "wow!" moment for customers).
- **Invest in influencer marketing. Partner with influencers at all levels and experiment with creative formats to boost brand awareness.** Supplement this with brand marketing and symmetric messaging across channels.

- **Amp up traditional digital marketing.** Optimize content marketing and performance marketing to augment organic growth.
- **Experiment with sales-assisted strategies.** For B2B products, build a digital-first sales journey and involve account reps at key points.
- **Explore co-branded partnerships.** Partner with other brands to attract new customers at lower costs than traditional marketing efforts. (See Figure 10.2 for popular examples.)

FIGURE 10.2
Co-branded partnerships could include one-time discounts or product collaborations or span into long-term strategies with embedded experiences.

- **Expand to new distribution channels.** Configure new package and pricing options that support new distribution channels.
- **Reframe the offering for a new market.** If other markets can benefit from similar functionality, reposition the offering and expand into a new category.

LEARN MORE

GROWTH PRACTICES

Product growth techniques require expertise across marketing, product, data strategy, and experimentation. To cultivate your toolkit, start with these resources:

- "Elena's Growth Scoop" on Substack by Elena Verna
- "Behind Product Lines" on Substack by Aatir Abdul Rauf
- *Hacking Growth: How Today's Fastest-Growing Companies Drive Breakout Success* by Sean Ellis and Morgan Brown
- *The Product-Led Playbook: How to Unlock Self-Serve Revenue and Dominate Your Market (with a Tiny Team)* by Wes Bush

Managing Scale and Optimization

How can I prepare for scale?

When organic growth takes off, you'll discover how scalable your product and operating model truly are. If you weren't serious about scale readiness and operational excellence in early iterations, systems and processes will struggle to keep pace with an influx of new customers. Of course, you'll be so busy with bug reports, customer complaints, and creating one-off updates, you'll miss the ugly truth: your product and team model are not ready for scale.

Attempting to scale when you aren't ready puts the team on a dangerous path that ends in excessive fire-fighting each day and less time working on high-value enhancements. (Trust me, I once watched teams of engineers spend nine months fixing latency issues because our #1 user complaint was app speed. Not a good look for our company, and the work wasn't exactly thrilling, either.)

Fortunately, you can avoid chaos and protect everyone's sanity before the mess happens. (See Figure 10.3 for your stress-free scaling checklist.) Scaling well isn't just the PM's responsibility—the entire team will need a mindset shift. If quality standards were low when the team expected 100 customers, that mindset won't scale to thousands or millions of active users. To navigate the change curve, the entire team must improve quality and efficiency across the board. If you're feeling stuck, review the checklist as a team, agree on your strategy, and publish the changes. And then stick to the plan!

The PM's Checklist for Stress-Free Scaling

- [ ] Confirm product-market fit and a path toward sustainable value.
- [ ] Ensure ROI makes financial sense and secure new funding or head count.
- [ ] Update GTM plan with new phases, tactics, and milestones.
- [ ] Update product roadmap, strategy, and backlog to reflect next steps.
- [ ] Review and update ways of working with the team to drive clarity and speed.
- [ ] Address bugs and defects consistently to avoid scaling known problems.
- [ ] Assess system quality, recoverability, and performance and future-proof gaps.
- [ ] Assess tech debt levels and address major concerns before scaling.
- [ ] Update data strategy, including data model, data governance, analytics, and reporting tools.
- [ ] Optimize internal processes to handle an influx of customers, using automation when possible.
- [ ] Allocate development capacity (ideally 20%) for ongoing production support and bug fixes.
- [ ] Establish consistent checkpoints and product reviews to evaluate wins, misses, and risks and overall performance.

FIGURE 10.3
Protect your peace and brand by preparing for scale early.

Operating a Product at Scale

When managing a product at scale, it's common to see product strategy pillars that promise to *optimize customer experience* or *increase operational efficiency*. These are fancy ways of saying the company wants to extract more value from this product. I'll cover optimization in the next section, but for now let's focus on *operational efficiency*.

When a strategy requires operational efficiency, you must find ways that maximize a product's value by reducing costs and increasing productivity. Start by conducting value chain analysis across critical customer journeys to uncover slow or inefficient processes that drive up expenses for the solution and customer friction. Then complete process-mapping exercises across internal workflows to understand where slowdowns occur in the product development process. Across both exercises, take note of repetitive or tedious tasks and areas that require multiple handoffs.

Both exercises will produce opportunities to solve via automation tools, platform improvements, or emerging technologies. And if you're starting to wonder about AI-powered technologies, your instincts are spot-on. AI solutions are excellent at solving productivity issues at scale. Figure 10.4 shows industry-agnostic ways you can use AI solutions to boost efficiency for customers and your team.

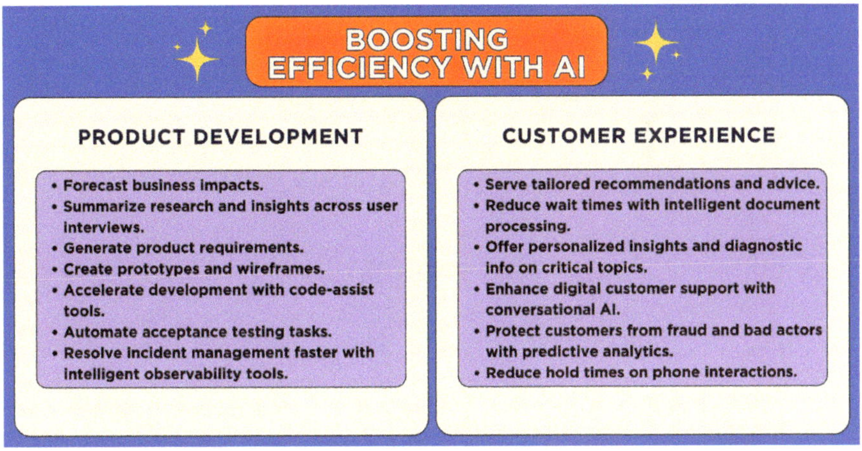

FIGURE 10.4

Speed up time-consuming processes for your team and your customers using AI.

PLAY 20

Optimizing with A/B Tests

How can we assess the impact of one experience over another?

Once the product has a sizeable user base (think thousands or millions of users), you'll unlock a new level in your product journey: enough volume to run in-market experiments! I'm biased, but I think this is the most exciting part of product management. Experimentation gives you freedom to learn and feel confident in your product choices, and it also eliminates some of the daily pressure to be right all the time. (Honestly, are we ever truly right?)

In-market testing can guide loads of choices, including growth tactics, conversion strategies, UX improvements, and new feature roll-outs. Techniques include A/B testing, multivariate testing, and experiments powered by machine-learning algorithms. Here's a breakdown of how they differ:

- **A/B testing** is when two or more variants of an experience are shown to users at random and statistical analysis indicates which variation performed better against a specific goal. This testing approach is commonly used to experiment with layout, content, feature rollouts, and other use cases shown in Figure 10.5. This form of testing is the easiest to design, implement, and analyze and commonly used to answer questions like "What landing page converts better?" "How does color affect our click-through rate?" or "How will this new feature impact 30-day activation?"

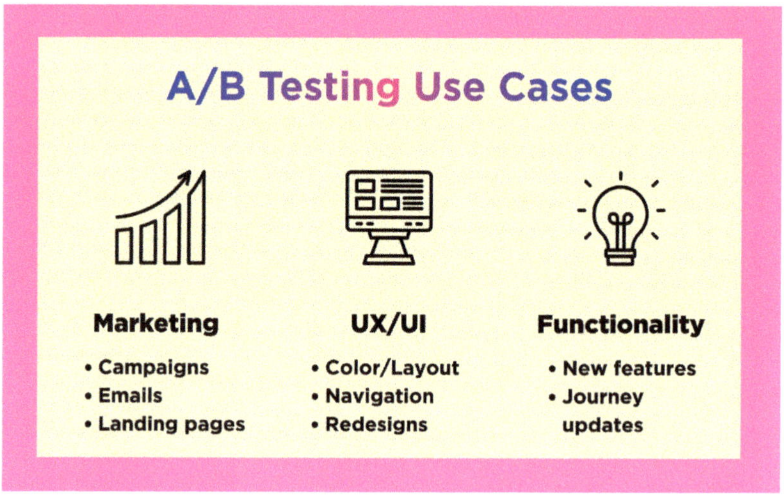

FIGURE 10.5
New to experimentation? Try A/B testing first.

- **Multivariate testing** is when different combinations of multiple elements are shown to users at random, and statistical analysis indicates the optimal combination. This testing method works well when testing the interaction, and when changing text and visual elements at the same time.
- **Multi-armed is a form of A/B testing** powered by machine learning algorithms that auto-allocate traffic to the winning

treatment while the test is running. This approach is great for high-traffic, short-duration experiences, like news headlines or seasonal promotions. Contextual bandit testing takes this one step further by leveraging machine learning algorithms that use customer context (i.e., age, time of day, location) to identify the best treatment for the specific user. This form of testing works well across highly-dynamic, personalized experiences.

If you're new to in-market experimentation, learn to master A/B test design first. Of course, you'll need a set of capabilities to launch in-market tests, so check with your product and engineering partners first to understand what tools are available to design, deliver, and analyze tests in production. Many companies integrate with third-party solutions, such as Optimizely, Statsig, or Adobe Target, to power their experimentation journey. That said, if you have a clear problem to tackle, the capabilities to launch a test, and an open mind, you're ready to dive into experiment design!

Use When

You want to assess the impact of one experience over another using in-market data.

Average Time

1 day to build the test plan, plus time to develop the test, launch it, and collect results

Try This

1. **Specify the problem statement and a testable hypothesis.**

   The first step in any experiment is to start with a clear problem statement and a testable hypothesis you can deliver in-market. Follow the madlibs below to create structure for your test. If you have enough resources and customer volume, consider testing two or three hypotheses against the control to maximize insights.

   - **Problem statement:** We've observed [X], which is causing [Y reaction] for [Z target user]. Our data suggests this problem is affecting [# of people] and costing our business [$ dollar value]. How might we [build, change, create the intended experience] for [target user] so that [insert desired outcome]?
   - **Hypothesis:** We believe that [doing X] will result in [desired outcome] for [target customer].

2. **Define success metrics and kill metrics.**

   When picking success metrics, choose a metric you can measure across all cohorts in the test, including the control. If you compare each cohort using different metrics, the results will be inherently flawed. Pick a success metric that most closely represents solving the problem (not a leading indicator or a micro-conversion.) For example, if you're launching a test in the sales funnel, your success metric should be customer conversion rate, even if the test experience happens at the top of the funnel. And don't forget to include kill metrics! Plenty of tests generate unintended, negative consequences; consider this in advance and ensure that you monitor these throughout the test.

   - **Success metrics:** If the test experience results in [X% increase/decrease] of [insert key metric here] compared to our control experience, we'll know we're successful at addressing the problem.
   - **Kill metrics:** We're not willing to sacrifice [this outcome/measurable impact], and we will stop the test if we observe it causing those impacts.

NOTE **DETERMINING SAMPLE SIZE AND TEST LENGTH**

Sample size represents the number of unique users you need in your control and test cohorts to generate reliable results. If you know the sample size for your test, you can project how long your test might take based on traffic to the solution.

To determine sample size, you'll need the current or baseline conversion rate for the event you're trying to influence, the minimum detectable effect or smallest lift you are willing to detect, and the statistical significance threshold you need to see to feel confident in the results (usually 95% or 99%). When you have these pieces of information, use Optimizely's free sample size calculator to determine your total sample population (**www.optimizely.com/sample-size-calculator**).

3. **Finalize and socialize the test plan.**

   Designing a strong experiment takes a great deal of intention and thorough planning. After Steps 1 and 2, document your test plan (as shown in Figure 10.6) and socialize it to avoid test collisions or unintended consequences. Every test plan should include:

- Problem statement, hypotheses, success metrics, and kill metrics
- Mock-ups or screen recordings of the control and test experiences
- Eligibility and suppression criteria that explain who should or shouldn't be in the test sample
- Test split percentage that represents how much traffic you want in each cohort
- Data and analytics requirements, including data you must collect and how to view it
- Mitigation techniques you'll use if the test negatively affects customers or the business

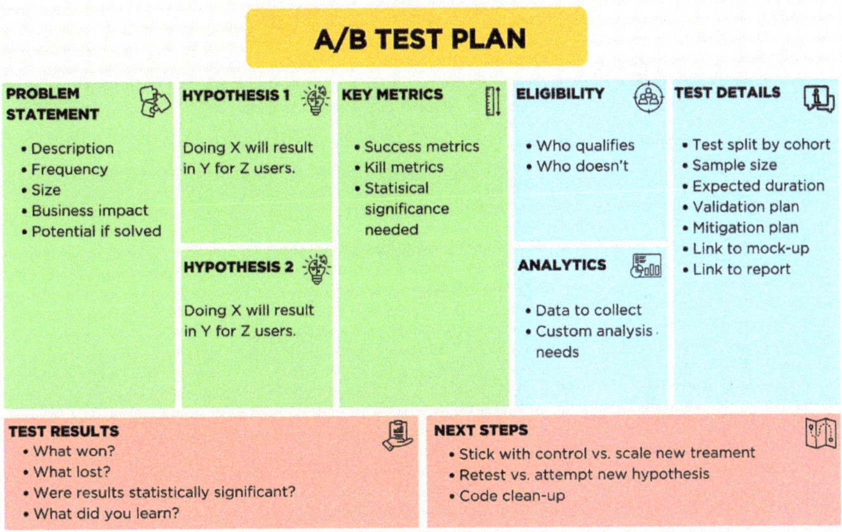

FIGURE 10.6
Use this structured template to socialize the hypothesis, implementation details, results, and next steps for each test.

4. **Design, launch, and check the experiment.**

   When you feel good about the test plan, start developing the test. Unless you are using a self-service A/B testing tool, work with design, engineering, and analytics to create the experiment.

   Before launching, perform regression testing to make sure that you didn't miss scenarios that affect test results. When launching the test, start by throttling a low amount of traffic into the test experience and confirming that things look right. For example, evaluate your kill metrics and technical performance metrics in each variation to ensure that issues like latency and errors aren't biasing results.

Once you're confident the experiment is working properly, increase volume into the test. In some cases, you may need to wait for data to normalize over time. For example, if you're launching a test on a marketing landing page, you can probably get an instant read on performance. But if you've just introduced a new hero feature for half of your customers, it's best to track feature adoption over a longer timeframe to ensure that shiny object syndrome doesn't bias your results.

5. **Collect and analyze results.**

   The instant you launch a test, start tracking results. If you're using an off-the-shelf testing tool like Adobe Target, Statsig, or Optimizely, you'll be able to view test results quickly in their self-service UI. In some cases, you should request custom analysis to analyze test data across proprietary business metrics or cross-channel cohorts. Ultimately, strong test analysis answers the following questions:

   - What experience won and what lost?
   - What metrics justify this conclusion?
   - Were the results statistically significant?
   - Was anything inconclusive?
   - What did we learn from this test?

6. **Determine next steps and share insights!**

   After analyzing results, it's time to finalize next steps and share out. If your test hypothesis resulted in a positive impact, develop a path to scale. If your test hypothesis was flat or only performed marginally better than the control, look for signals that might explain why and consider testing a new hypothesis before making a permanent change.

   On the other hand, if your hypothesis loses against the control experience, you haven't failed. In fact, most tests fail! Just make sure that you learn from the results and share insights with others. To make it fun, turn your test results into a quiz or share them with GuessTheTest (see Figure 10.7).

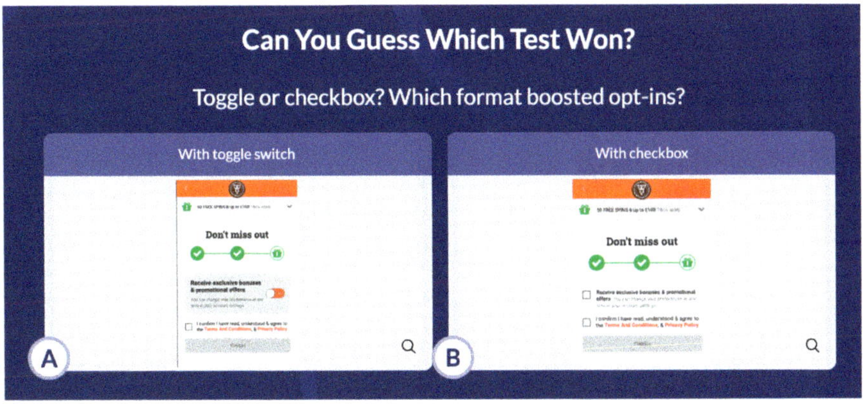

IMAGE COURTESY OF GUESSTHETEST

FIGURE 10.7

GuessTheTest is chock-full of A/B test case studies and results to inspire you.

LEARN MORE

BEEFING UP YOUR EXPERIMENTATION MUSCLE

The digital experimentation community is a gold mine for those seeking A/B inspiration and guidance. If you want to learn more about statistics and testing methods, check out these resources:

- ***You Should Test That: Conversion Optimization for More Leads, Sales, and Profit or The Art and Science of Optimized Marketing*** **by Chris Goward**
- **Blogs, white papers, and sample size calculators by Optimizely, Statsig, VWO, or Adobe Target**

Sunsetting Products

Despite your best attempts to manage a product, sometimes the best path forward is to sunset the solution. No, this type of work doesn't scream "razzle-dazzle," but you should feel immense pride when killing a product that eats up company resources for little in return. Seriously, wear it like a badge of pride!

In many cases, deprecating a product becomes a replatforming, refactoring, or migration effort because the product itself still delivers business value, but the technology leaders determined the solution architecture is so far behind that it's harming the business. However, some companies struggle to sunset solutions because retiring a solution involves brand risk, or they're too busy pursuing "shiny new objects" and don't have the strategic maturity to sunset solutions proactively. If you're wondering whether your product is ripe for deprecation, review the signs in Figure 10.8. If these resonate, then your product is in decline and needs an official path toward deprecation.

FIGURE 10.8
When you see these signs, establish a new direction and deprecation strategy.

If you believe your product or feature is a strong candidate for deprecation, take these steps to get it started and lead the way forward:

1. **Form a compelling business case for deprecation.** To sunset a solution, leaders need a clear business case before assigning resources to the project. Map out the reasons for deprecation, risks and consequences of doing so, options to overcome risks, and expected benefits in return.
2. **Determine the new strategy.** Every deprecation strategy must include an offboarding journey for current users. Your offboarding strategy should include functionality requirements to help users save or export data, in-app messaging to keep customers informed, and a migration plan when you're directing users to a new solution. If you're building a new solution in parallel, document and prioritize user needs, craft a revised value proposition, and identify requirements for the new product, too.

3. **Complete a technical feasibility assessment.** Partner with the engineering team and follow the steps in Play 10 (Chapter 6, "How: Defining the Approach"). If the company can't commit ongoing resources to support another solution, consider buying a solution instead.
4. **Align on the path forward and begin deprecation.** Craft a vision, strategic pillars, and a roadmap with phased cut-off dates using methods in Chapter 7, "Establishing a Plan and Getting Buy-In." After securing buy-in and resources, prep the backlog, host a kickoff, and start the work.
5. **Emphasize user communication during deprecation process.** Users need to know what to expect, when to expect it, and if workarounds will exist after you eliminate functionality. Keep them informed by using emails and in-app messages well in advance. For example, before sunsetting Skype in 2025, Microsoft used clear in-app messaging and allowed users to maintain account credentials, contacts, and chat history when transitioning to Microsoft Teams. (See Figure 10.9.)

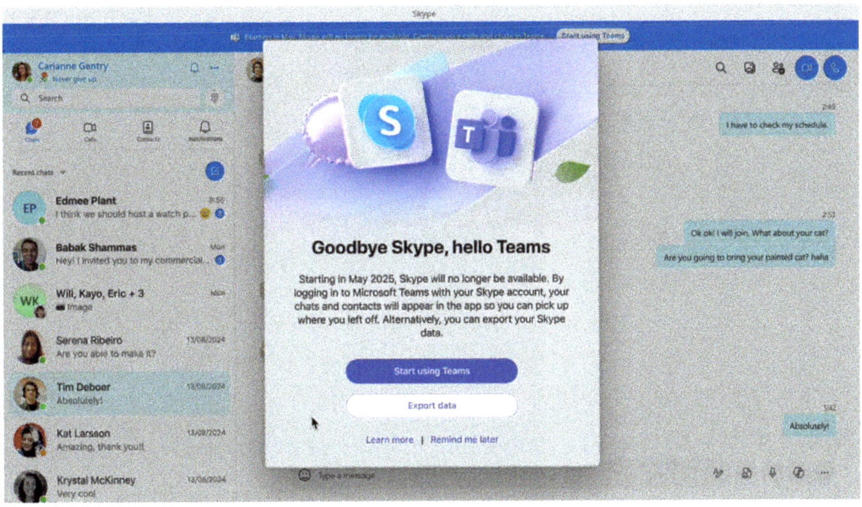

IMAGE COURTESY OF MICROSOFT

FIGURE 10.9
Sunsetting a product can be elegant and thoughtful.

PLAY 21

Rehabbing a Fledgling Product

What should I do when my solution isn't driving value or accomplishing goals?

Most PMs don't spend their entire careers supporting glamorous product in hypergrowth mode or beloved solutions like Google Maps, Slack, or Netflix. Thousands of PMs work in the trenches, supporting unglamourous products and systems that need tender loving care.

So, at some point, you'll inherit an existing product, feature, or platform that stinks. The documentation will be scant, the feature set will confuse you, and the development team will complain that no one listened to their concerns years ago. You can be irritated for about five minutes, but then you need to move on.

In fact, when you encounter a product that's sinking, think of it as a gift. It's a chance to help a team in need, drive clarity for the company, and include a new résumé bullet focused on product turnarounds and team transformation. (That kind of story is perfect for your next interview!) Here's how to right a sinking ship.

Use When

You've inherited a solution that isn't driving value or accomplishing its goals.

Average Time

1 week, plus time to implement recommendations

Try This

1. **Assess current state using quant and qual data.**

   Anytime you begin a new role, start gathering data to assess how your product, platform, or feature is performing. While it's tempting to outsource this work and simply ask others for their opinion, it's critical that *you* conduct this analysis. Doing so helps you establish a stronger point of view on the product and discover reporting gaps. To assess your product, try the following:

   - **Assess current product performance with quantitative data.** Evaluate customer success, business success, and product quality using the metrics cited earlier in this chapter. Use clickstream tools, funnel reports, observability tools, and internal dashboards to understand what's happening right now and what the annual trends were over the last two years.

- **Put yourself in the customers' shoes.** Identify customer friction hot spots by reviewing session replays, parsing through customer complaints, and testing out the product yourself. Listen to customer support calls, check ratings and reviews, and get testing credentials to experience the product just like the customer does.
- **Get feedback from the team.** After doing your own digging, gather feedback from the team about what's working well, what's not, and how they would improve the product. If they offer recommendations, ask them for the specific metric or business outcome the change would improve. During this step, be prepared to gather insights about the product strategy, the team culture, the execution process, and everything in between.

After evaluating the data, synthesize the feedback and themes into a readout for yourself. Keep a running list of product insights and feedback as you gather more information. For some, it's helpful to bucket the data into the same groupings shown earlier in this chapter under "Monitoring Product Performance."

2. **Confirm the business goals.**

If you're inheriting a years-old solution, business goals may no longer be the same. Historical context is always important, but you need to confirm what success means *now*. Identify your key decision-makers, executive sponsors, and partners and pull them together to answer the following:

- What does business success look like for this product right now? (For example, growth, retention, cost savings, revenue, etc.)
- What are the specific metrics that matter most over the next 1, 3 and 5 years? (Force rank when possible.)
- Who is the ideal customer or user for this solution and why do we care about them?
- Have any new triggers, data, or assumptions emerged that suggest we should revisit the core product vision, product strategy, or critical choices about technology and product design?

3. **Develop your point-of view.**

   After completing Steps 1 and 2, start synthesizing what you heard. Document suboptimized journeys, gaps in product quality, or misalignment on the foundational hypotheses that form the Product-Market Fit Pyramid. Then break the insights and recommendations into two buckets: tactics to optimize the current state solution and strategic decisions needed for long-term success. The buckets might look like this:

   - **Quick wins to optimize the solution right now:**
     - Quality and usability improvements, including defect reduction efforts, content updates, and error messaging improvements
     - Small enhancements to improve key metrics across acquisition, onboarding, activation, or retention journeys
     - A/B test proposals to learn in-market or analytics
     - Observability improvements to improve telemetry on metrics
   - **Strategic decisions to improve long-term success:**
     - Reframe business goals into objectives and key results.
     - Establish customer segments or reassess target user needs.
     - Conduct feature set analysis using importance-satisfaction surveys.
     - Update the product vision and strategy.
     - Refresh pricing strategy.
     - Reassess technology needs, including business capabilities and architecture plan.
     - Establish clear design principles and conduct evaluative research to improve product design.

4. **Begin dual-track execution.**

   If you have development resources waiting for direction, review your list of tactical recommendations with them and add them to the team backlog in priority order. The team should knock these out while you begin refreshing the product vision, strategy, and roadmap.

While the team delivers tactical improvements, add a discovery swim lane to the product roadmap and fill it with time-bound activities that lead to a new vision, strategy, and roadmap. (Use the cheat sheet in Table 10.5 to help.) Depending on the amount of work required, this phase could take 2–3 months or longer, but don't start making decisions single-handedly. Be inclusive with design, tech, and business partners to increase alignment. Personal heroics don't rescue products; high-performing teams do.

TABLE 10.5 THE PRODUCT MANAGER'S CHEAT SHEET TO ESSENTIAL PLAYBOOK METHODS!

If You Need to Revisit . . .	Try This . . .
Business goals and strategic assumptions	Business strategy workshop (Chapter 3)
Target customer and user needs	Generative customer research (Chapter 4)
Product vision and value proposition	Evaluative customer research (Chapter 5)
Core functionality and feature set prioritization	Build a problem-solution fit map (Chapter 5)
Design choices and assumptions	Usability testing, experience optimization (Chapters 5, 10)
Technology choices and assumptions	Capability upgrades, quality improvements (Chapters 6, 10)
Business projections and key metrics	ROI analysis (Chapter 7)
Pricing, distribution, or marketing decisions	Review go-to-market details, pricing and marketing strategy (Chapter 9)

Pulling It All Together

Have you ever hiked a mountain with friends? The closer you get to the summit, the group's energy ebbs and flows, you hear complaints about gear, and everyone's wondering how much longer until the view.

Managing an in-market product is no different. You're managing the team's energy and focus as they tackle new feature development, optimization, and production support all at once. The system architecture and operating model might reveal a few cracks, but it's still getting the job done. And there's always someone a little impatient who wants to know when the next big release is.

But guess what? Despite the day-to-day humdrum, you're still doing something incredible and brave. You're climbing a mountain under uncertain conditions, you're adapting to unique personalities and demands, and you've created something in the world that was simply a mere idea at one point. Even if you aren't at the summit yet, it's okay to appreciate the view right from where you're standing. It's still impressive.

CHAPTER 11

Mastering Leadership and Influence

Early in my career, I landed a digital product management gig at a major banking institution. Because our products supported millions of customers, leading one sprint team was the official standard for every product manager. Leading *two* sprint teams, however, was the unofficial requirement for promotion.

Leading my first sprint team felt like a warm blanket. Delivery velocity was healthy, ceremonies were productive, and the team genuinely enjoyed working together. We had the memes and laughter to prove it. When I expanded to a second team, the contrast was stark. Meetings were quiet, quality was inconsistent, and the tech lead was abrasive. During team ceremonies, he'd interrupt me, impatiently wait for the meeting to end, and rush off afterward.

The situation puzzled me. *How could two teams be so different? Why was the second team underperforming? Had I done something to irritate the tech lead?*

A few years prior, I'd completed conflict resolution training, so I decided to put the methods to use. I invited the tech lead to meet in a quiet space and calmly described how I felt sidelined during team meetings without blaming or accusing him. I said that I wanted to improve teamwide communication and my first step was to start between us. Then I asked him to describe his experience. Within a few minutes, I learned he was struggling, too. That year, a series of back-to-back losses saw two PMs join and swiftly depart the team. He tried to keep work flowing, but without a dedicated PM, quality suffered. To make matters worse, his boss piled on the responsibility of managing a second team of engineers. Suddenly, it clicked. He was burning out.

During that conversation, we formed empathy for each other and committed to creating a respectful team environment. After our chat, the heaviness lifted. Standups were calm, communication was respectful, and we both started building a supportive team culture.

But, if you think I was perfectly calm before that conversation, allow me to clarify. My heart was racing, palms were sweaty, and I imagined at least ten different ways the conversation could go sideways... and guess what? Our talk didn't go sideways—it went *forward*. And it taught me a valuable lesson I carry each day: true leadership requires courage.

As a product manager, your daily mission is daunting: motivate and influence dozens of partners to deliver successful outcomes for customers and the company. Some partnerships are smooth, and some are challenging. To build trust and influence, you'll need to be a coach, cheerleader, armchair therapist, or simply a mirror. Some days, the task is trickier than others and anxiety creeps in. When you notice that happening, push yourself to lead with courage and ask the questions that matter most:

- How can I build a high-performing team?
- How can I forge stronger alignment with partners?
- Am I growing as a leader?

Building a High-Performing Team

Most product managers work with a dedicated group of cross-functional partners to design, ship, and optimize products. Whether you call it a pizza team, a product team, a pod, a sprint team, or something else, these are the folks in the trenches with you attending daily standups, conducting research, designing the solution, and building the product. This is your *core team.*

In an ideal world, everyone on the core team trusts each other, collaborates effectively, and meets expectations, so you can spend more time working toward goals and less time battling interpersonal issues. But mature, high-performing teams are not accidental—they are intentional. They form over time by consistently modeling trust, clarity, and healthy team dynamics.

As the product manager of a team, it's your responsibility to cultivate a high-performing team that delivers positive outcomes. But your team leader role comes with a twist! You have zero formal authority on the team, and you don't get to pick your teammates. Instead of relying on an org chart and reporting relationships to influence the group, you must embody leadership behaviors that nurture trust, respect, and intrinsic motivation. In other words, you need to think like a *player and a coach.*

As a player and coach on your core team, you have dual responsibilities. In the player role, your first responsibility is to effectively execute tasks that fuel clarity and set direction for the team. These traditional product management tasks are building the product vision, strategy, and roadmap; representing the voice of the customer; and ensuring that the team is focused on high-value problems and solutions. As the coach, however, your responsibility is to prioritize the team's health and success by modeling and cultivating strong performance. In this capacity, you'll shift focus from *what* the team is doing to *how* the team gets it done.

Most of this book already covers methods to be successful in the player/product manager role, but to be an effective coach/team leader, you must practice behaviors that shape and increase trust, motivation and growth for everyone on your team. Start by embracing five key leadership behaviors each day:

- Model integrity
- Prioritize team growth and adaptability
- Value excellence
- Handle difficult situations with maturity
- Follow the golden rule

Model Integrity

Integrity and trust are the foundation of high-performing teams. In the player-coach concept, the product manager embodies honesty, consistency, and fairness in their actions and words. The product manager doesn't commit to dates or estimates without consulting the team, doesn't compromise on team values, and accurately represents trade-offs, risks, and status updates. The team trusts the product manager to make fair decisions and communicate truthfully and consistently. In turn, the product manager expects the same integrity and accountability from team members.

Prioritize Team Growth and Adaptability

The product manager encourages team growth by focusing on team dynamics proactively and hosts fruitful retros to help the team reflect and collaborate on opportunities (Figure 11.1). They also engage peers to change processes or roles and improve ways of working. When a reorg, leadership reset, or strategic pivot occurs, the PM helps the team adapt.

IMAGE COURTESY OF MIRO

FIGURE 11.1
Try the Sailboat retrospective technique to run healthy retros.

Value Excellence

The product manager takes pride in delivering high-quality products by practicing excellence in their field. They follow and communicate learnings from thought leaders, competitors, industry benchmarks, and mentors. They encourage cross-functional peers to do the same and share rationale that explains the why behind business, technology, data, and product design choices.

Handle Difficult Situations with Maturity

Product managers understand and accept their responsibility to lead in difficult situations. During stressful situations (see Figure 11.2), the product manager stays engaged, level-headed, and leads with facts until a solution is reached. Afterward, they evaluate root-cause with the team, offer constructive feedback, and accept accountability for their mistakes. They do not dwell on failure or blame, nor do they allow their teammates to do so.

FIGURE 11.2
Difficult situations happen. Strong PMs stay calm under pressure.

Follow the Golden Rule

The product manager always treats the team with respect and consideration. They don't waste the team's time, expect the team to pick up their slack, or behave disrespectfully toward members of the team. They intervene and coach when others act in a way that damages relationships on the team. When appropriate, they escalate issues to protect the team's psychological safety.

Recognizing Team Contributions

Recognizing others is so important that it deserves its own section. Practicing team recognition is one of the easiest ways to reinforce positive behaviors and strengthen connections on a core team. When colleagues feel seen and appreciated, they're motivated to put their best foot forward. On the other hand, when your team is working extra hours to meet a critical milestone and no one stops to say thanks, morale tanks and resentment builds. Just like you, *everyone* wants to feel valued at work.

The good news is that it only takes five minutes to show appreciation and there's *always* something to appreciate. Think back to your last sprint. Did someone develop a new skill, demonstrate initiative on a customer complaint, or lend a hand to a colleague in need? If so, then celebrate their superpower and progress! Shining the spotlight on your teammates improves their visibility with executives and reinforces a culture of appreciation and respect across roles. (See Figure 11.3 for ideas and inspiration.)

FIGURE 11.3
My personal favorite is the haiku!

To those who rarely celebrate team progress, *stop being chintzy with recognition!* When you only recognize people once a quarter or when the team hits a specific target, you miss an opportunity to build a culture of continuous encouragement. It's much easier to celebrate when performance looks great, but sometimes people need an extra bit of encouragement in the dead-middle of a months-long re-platforming effort or when they are struggling to unwind a complex security issue.

Resolving Conflicts

Being a team leader comes with light and shade. Some days, you'll celebrate and uplift the team. Other days, you'll need courage to confront uncomfortable issues. To be an effective team leader, you must lean into all the shades of leadership, including moments that require conflict resolution.

Conflict is a natural outcome any time you assemble a group of people with unique backgrounds, personalities, opinions, and expertise. The presence of conflict isn't necessarily a bad thing, either. In fact, disagreement on high-performing teams might indicate that the team is engaged and comfortable sharing different opinions, which can lead to stronger solutions. However, when the conflict threatens trust, communication, and clarity on the team, it's important to address it directly. If you see conflict on the team, and you're unsure if it's helping or harming the team, ask yourself these three questions:

1. Is this conflict helping the team explore different perspectives and make better decisions?
2. Is this conflict helping the team move toward their goals?
3. Is this conflict helping to build productive work relationships that enhance trust?

If the answer is yes to all three questions, the conflict is likely healthy. If you can't say yes to all three questions, the conflict could be unhealthy and require your intervention.

Conflict Resolution Techniques

Addressing conflict is one of the most uncomfortable skills to develop as a team leader or junior PM. As humans, we shy away from conflict because we fear how others will react, or we don't have skills to handle it effectively. But learning to address unhealthy tension now will give you courage and confidence to handle future conflicts that arise in the workplace (and yes, they *will* happen).

As a team leader, it's important to step back and consider why conflict emerges on teams. It's normal to see conflict arise when the team experiences a shift in identity, composition, mission, or size. For example, org changes or strategic pivots often result in some conflict or tension on teams. If you believe this type of change is causing team confusion or conflict, you can research Tuckman's Five Stages of Group Development (see Figure 11.4) to assess the operational maturity of your team, diagnose what stage you're in, and identify tactics to help the team shift into high performance mode.[1]

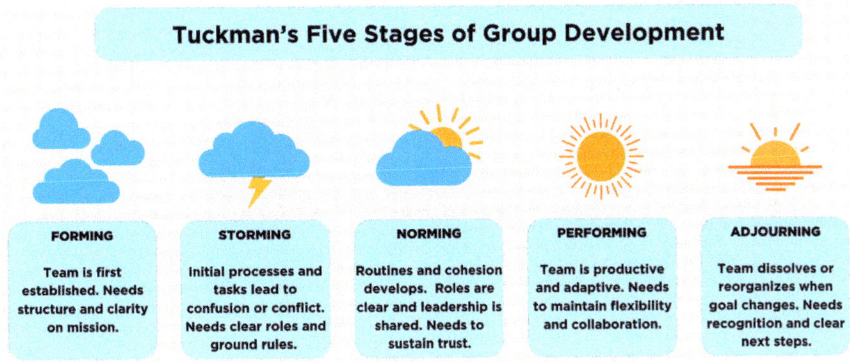

FIGURE 11.4
Performing is the ideal stage for core teams.

Conflict also occurs when individual styles clash in a group. When you see interpersonal conflicts arise on your team, review the Thomas-Kilmann Conflict Mode Instrument (see Figure 11.5) to assess the situation and identify unproductive patterns on the team.[2] This model highlights five different ways that individuals handle conflict in a group setting and can help members of the team increase self-awareness when conflict occurs.

---

1 This was originally published in 1965 as a four-stage developmental sequence to describe small group development. In 1977, Bruce Tuckman added a fifth stage, "adjourning." Today, many leaders use this framework to evaluate team maturity and build action plans. To learn more about Tuckman's Five Stages of Group Development, visit www.wcupa.edu/coral/tuckmanStagesGroupDelvelopment.aspx

2 Developed in the 1970s by Kenneth W. Thomas and Ralph H. Kilmann, the Thomas-Kilmann Conflict Mode Instrument originated as a research tool to analyze and assess conflict styles. Today, team leaders can use this model to facilitate dialogue with teams or conduct a formal assessment of individual and collective conflict styles. For more information, check out https://kilmanndiagnostics.com/

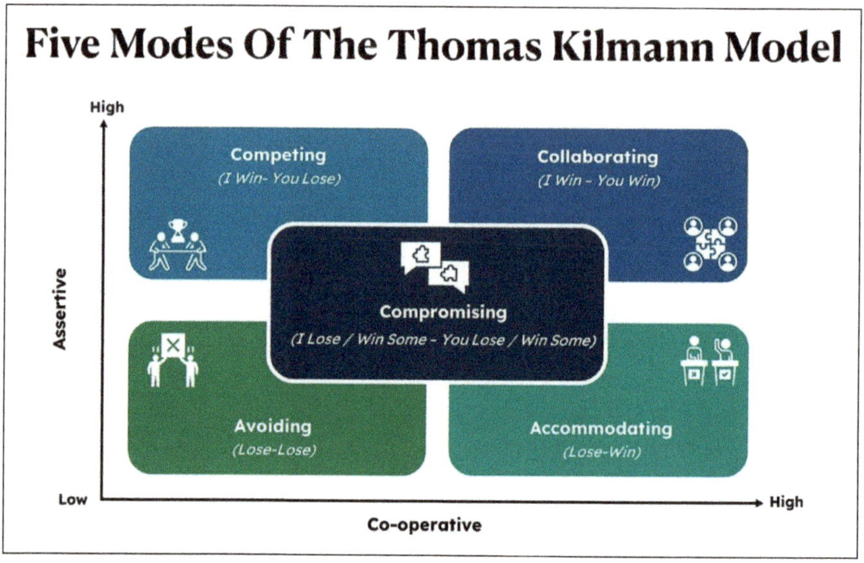

IMAGE COURTESY OF KAPABLE

FIGURE 11.5
Share this model with your team to increase self-awareness and align on the ideal way to handle conflict.

Sometimes, the conflict is between you and another person. When that happens, it's important to address the issue directly in a way that promotes fairness, shared understanding, and a solution-focused mindset. The book *Crucial Conversations* (see Figure 11.6) highlights a three-part method that helps you prepare for tough conversations, host a productive discussion, and agree on future actions to eliminate conflict. As a team leader, I use this technique to host performance conversations with direct reports, tackle uncomfortable tension with partners, and coach dozens of product managers to resolve conflicts before they escalate.

FIGURE 11.6
Buy the book, find the courage, hold the conversation. You won't regret it.

Increasing Alignment with Stakeholders

As the primary evangelist for their solution, product managers spend a significant amount of time engaging stakeholders *outside* of their core team, too. On any given day, a PM might interact with marketing experts, customer support reps, pricing analysts, data partners, business development teams, and of course, senior leaders.

Each partner is vital to the product's success, but in many cases you'll interact less frequently with them than with your core team. That means each engagement and interaction you have with them must be highly effective at influencing and aligning them to a goal. To strengthen alignment and build strong bridges with stakeholders, follow these techniques:

- Align on expectations.
- Master concise storytelling.
- Establish consistent, fair processes.
- Learn to say no.
- Practice mutuality, not reciprocity.

Align on Expectations

Every stakeholder or partner needs clarity on their role and responsibility within your engagement, but not all partners need the same level of attention or have the same decision-making power. If you or your partner is struggling with basic questions like ownership, accountability, and engagement levels during a project, co-create a RACI matrix or stakeholder map with your partner to clarify their involvement and contributions. (Or better yet, do this at the start of engagements to prevent confusion later.)

Master Concise Storytelling

When stakeholders aren't 100% focused on your initiative, they don't have time to read every customer insight, product requirements, or your latest release notes. Instead, communicate in a way that is concise, memorable, and highlights what matters to them. For example, use the following practices:

- When you're in front of executives, tie the product's story to big-picture outcomes, like market share, brand impact, competitive differentiation, customer growth, and profitability improvements.
- When meeting with marketing, talk in terms of customer benefits (not features).
- When engaging a dependent team for support, use the roadshow slides covered in Figure 7.6, Chapter 7, "Establishing a Plan and Getting Buy-In," to make the ask transparent and clear. Always include strategic context, the request, and a desired timeframe.
- When communicating a product update to stakeholders, consolidate the update into short bullets that cover the essential points, including a metrics update, recent wins or misses, new customer insights, and what to expect over the next 30 days. (See Figure 11.7.)

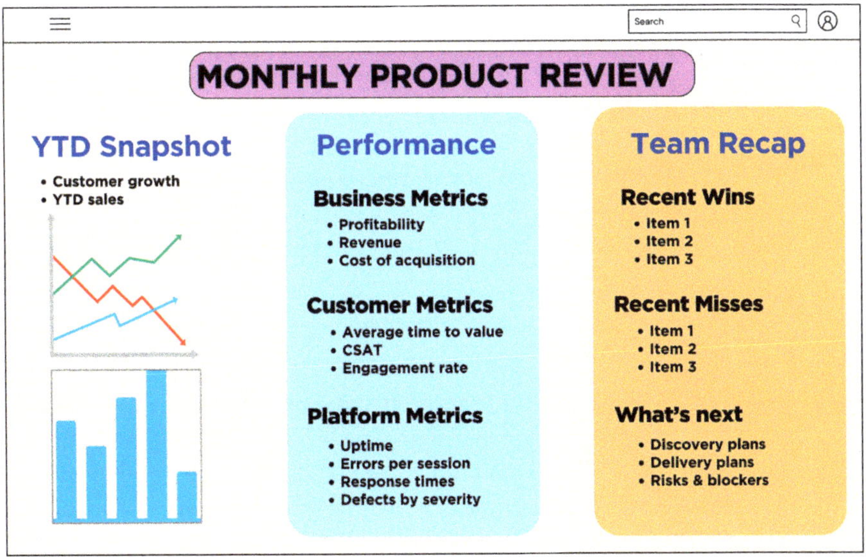

FIGURE 11.7

Master the art of the one-slider. It's concise, efficient, and repeatable each month.

Establish Consistent, Fair Processes

When your product is in growth mode or operating at scale, you'll notice a lot more interruptions and ad-hoc requests. (This is especially common on platform teams that have many stakeholders competing for shared resources or support.) In this scenario, stakeholders deserve clarity on how you operate and manage competing needs so they can plan accordingly. Instead of managing relationships one-on-one, try to design systems and processes that increase predictability, fairness, and transparency. For example, your partners should have a clear way to understand how you accomplish the following tasks:

- Prioritize new business requests.
- Conduct roadmap planning.
- Manage trade-offs.
- Communicate business and performance reviews.
- Manage production support and customer complaints.

Learn to Say No

Saying yes to every stakeholder request is a surefire way to lose credibility with your team, inundate the backlog with disjointed work, and start the journey of becoming a feature factory. That said, saying no to stakeholders is a delicate task because you need to preserve the relationship. When you receive a new feature request that doesn't align with customer or business priorities, here's how you can respond:

- **Start with thanks.** Every idea is a signal of support and enthusiasm for your customer or problem space. Thank them for considering a new way to drive value.
- **Lead with facts.** After saying thanks, present facts, not feelings. If you have evidence that suggests the request won't move the needle at all, or it won't move the needle as much as other candidates, share them. These can include business goals, performance data, tech estimates, customer feedback, competitor insights, or any other data points you use to prioritize work.
- **Tee up a trade-off.** If you're dealing with a senior stakeholder, consider framing the request as a trade-off. For example, you can say "To accomplish this request, I would need to delay our progress on [this critical priority] which we expect to deliver [this much value]. If you believe we should trade off this work for your request, let's discuss further. For now, I can add this to our parking lot of ideas for future consideration."
- **Look for win-win solutions.** Don't take every request literally. Oftentimes, the partner will throw out an idea or solution, but they really just want to fix a problem. Step back, think about the core problem the stakeholder wants to solve, and work the request from that angle. If you're aligned on the problem to solve, it's easier to find a win-win scenario that benefits customers and the company.

Practice Mutuality, Not Reciprocity

Strong partnerships are not built on reciprocity; they are built on mutuality. In other words, if your stakeholder relationships are transactional and focused on "give and take," they aren't built to last. Successful partnerships exist when both parties have skin in the game and a vested interest in one another's success. As a product manager, you can increase mutuality and strengthen stakeholder relationships by looking for opportunities to co-create ideas, giving credit to partners for their contributions and offering support when partners need it.

Leading as an Individual Contributor

Early in my career, a former boss asked a question that caught me off guard. "It costs two million dollars a year to staff your sprint team. How are you ensuring that the team is successful, and the company sees a return on that investment?" That question sparked an aha moment for me. On paper, no one reported to me, and I certainly didn't own a P&L. In fact, I sat on the lowest rung in the product management job ladder... but still, she expected me to think and act like a leader.

That interaction taught me that product leadership isn't an outcome that comes with a title of senior product manager, group product manager, or product VP. It's a mindset. While those roles may unlock people-management responsibilities and greater visibility, there's no better time to start developing your leadership style than when you're still an individual contributor. Growing where you're planted today helps you become a stronger leader tomorrow, but you need to be intentional about it (see Figure 11.8).

FIGURE 11.8
Chase growth over titles.

Develop Deep Product Sense

While many product management skills are transferable—like knowing how to tackle discovery, delivery, and leading teams—these alone are not sole drivers of success. Meaningful impact also requires a solid foundation of knowledge and keen judgment in your chosen domain. In other words, success relies on how well you develop *product sense.*

Product sense forms by fusing and synthesizing everything you know about the problem space— including customers, your industry, market dynamics, and competitors—and everything you know about the solution space—including your products, competitor solutions, technology and data trends, and what's on the horizon. It's directly tied to how much time you spend learning about factors that affect customers and your business.

Investing time to study and deepen domain knowledge sharpens your judgment over time and will make you a successful product leader. Believe it or not, tackling problem definition doesn't go away or get easier as you grow in leadership roles—it's quite the opposite! The problems get harder, and the risk of being wrong is more expensive, so you need deeper product sense to make the right calls.

Use Your Superpower to Lift Others Up

Every single person has a superpower, including you. I've seen junior product managers demonstrate wicked analytical skills, master partner integrations quickly, run top-notch user feedback sessions, triage bugs like a genius, and the list goes on! Many of them don't even realize how exceptional they are at a specific skill compared to their peers, so when they start asking questions about promotion, here's what I say: *If you want to be a product leader, you have to multiply your impact through others. How can you use your superpower to lift up the organization and those around you right now?*

We usually craft a plan within five minutes of asking that question, and you can, too. Once you identify a skill to teach, then crystallize your knowledge into a useful format and start sharing it. You can schedule a lunch-and-learn, build a template, host a training session, or create a recording. There are plenty of ways you can scale knowledge, but don't over-think it... just do it!

Don't Escalate Until You've Tried to Resolve the Issue

One of the most important skills in product leadership is knowing when to resolve an issue and when to escalate it. This is something early product managers struggle with because they conflate proactive communication with unnecessary escalations. It's common for junior product managers to notice a problem, get nervous, and share it with their leader right away by sending a simple message like this:

> "Hey, boss. Legal hasn't joined any of our product launch prep meetings. Thought you should know because this might affect our launch dates unless we get their sign-off soon."

A junior PM might assume sending a message like this proves they proactively share risks. But think about it from your leader's perspective: it seems like the PM is off-loading a vague problem before attempting to address it themselves. Think back to the concept of *extreme ownership*. A product manager with an ownership mindset handles a situation like this differently. When they spot a risk, they attempt to resolve it *before* escalating, and when the issue persists, their communication reads like this:

> "Hey, boss. Legal hasn't attended any of our product launch prep meetings or given their necessary sign-off, which puts our launch date at risk. In my experience, one-on-one meetings or a call between leaders is the most effective way to secure their sign-off. I've already attempted to host a one-on-one meeting twice, but the legal rep didn't join and hasn't offered to send a delegate. I would appreciate your feedback on attempting to schedule a call among leaders because I want to make sure that our team brand stays positive and the partnership is strong. Please let me know when you have a few minutes to chat about our options."

See the difference in the second example? By the time this product manager escalates and asks for assistance, their leader knows why the problem matters, what they've already attempted, and how to assist and support the PM. A message like this reveals higher leadership maturity.

Of course, you should ask for help when you're completely stumped. There will be times when you have zero context (this still happens to *everyone*, including senior leaders). But whenever possible, attempt to help yourself first and resolve before you escalate.

Practice Organizational Patience and Grit

Product management is a head-scratcher sometimes. Everyone's running around trying to demonstrate a bias for action and then suddenly you discover product development will take four times longer than you expected. And no, I'm not being dramatic, this stuff actually happens. Canva's CEO once explained how the company started a six-month re-platforming project that wound up taking two years. During that entire two-year time span, the company didn't ship a single feature.[3]

Canva's story might seem like an extreme example, but every company struggles with issues that can delay your dreams of shipping a world-class product on time. Sometimes, it's funding or resource constraints; other times, it's strategic misalignment or tricky dependencies. In some cases, the company starts a major transformation when you're in the middle of a years-long product initiative. When headwinds like this happen, you need to practice organizational patience and grit.

It's easy to become disengaged and negative when structural blockers throw your plans off track, but leadership is about finding a way to show up every day with resolve and a can-do attitude to help the team navigate stormy waters. (It once took me six months and seven attempts to secure resources for no-brainer work because the company's staffing model wasn't clear. Was it extremely frustrating? Yes. Did I give up? No. Did I ever get the resources approved? You bet I did...and that's true grit!)

Preparing for Product Leadership Roles

At a certain point, you'll reach the limit of what you can learn in a specific role and start pondering what's next. Some PMs pursue the individual contributor track and advance to roles that involve greater complexity and influence without people leadership duties. These people tend to become senior product managers, staff product managers, or even principal product managers. For others, they'll start to consider product leadership roles that involve people management responsibilities, with titles like senior manager of product

3 Lenny Rachitsky, "The Making of Canva," *Lenny's Newsletter*, November 2, 2025, www.lennysnewsletter.com/p/the-making-of-canva

management, group product managers, or directors of product. If you're considering the latter path, here's how to prepare for the next level.

Shifting from Product Manager to Product Leader

Product leadership starts with mastering the fundamentals because it's impossible to lead others well if you haven't done the work yourself. That's why most of this book covers tips and techniques to tackle the entire product life cycle and lead as an individual contributor (IC). Your experiences as an IC create the foundation for leadership roles and prepare you for the next big step: managing a much larger scope and sphere of influence as a people leader and coach.

As a product leader, you'll be responsible for managing *portfolios of products*, not just a specific problem space, product, or set of features. Instead of hands-on execution, you'll focus on aligning executives, dialing in the product vision and strategy, securing resources, and coaching product managers that work with core teams. In other words, your impact and scope is greater, but your success hinges on your ability to build strong teams and influence at scale.

The best way to prepare yourself for product leader roles is by acting like one, regardless of the title or salary you have right now. Figure 11.9 includes a set of development actions you can take to embrace a leader's mindset and start proving your mettle.

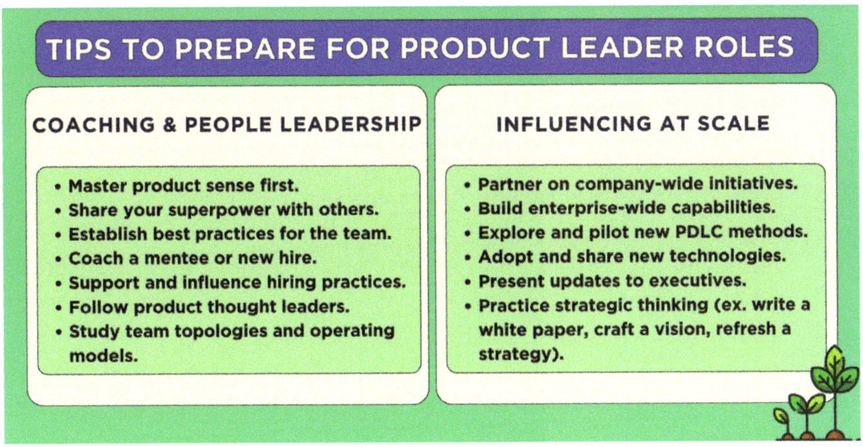

FIGURE 11.9

Pick a few goals from each category and bake them into your promotion-readiness plan.

State Your Intentions Early

Many PMs wait until their annual review conversation to express promotion desires and interest in people-management responsibilities. They believe that if they put in the effort and accomplish strong outcomes, their work speaks for itself. There are two problems with this approach:

- **Strong performance doesn't automatically lead to promotions.** Doing great work doesn't mean you're ready for the next level; it means you're doing a great job at your *current* level. Leadership roles require additional skills. That leads us to the next problem.
- **Leaders cannot coach or advocate for what they don't know.** If you don't communicate your promotion goals early, your leader can't structure work, coaching, and opportunities that prepare you for the next level. To get promoted, you need to demonstrate evidence that you're already operating at the next level, and your leader needs time (sometimes 6–12 months) to broadcast this evidence to peers and leaders. This is especially important in companies that use peer calibration and forced-ranking tools during review cycles.

Put simply, don't wait for someone to notice and promote you. Even if it scares you, have courage, advocate for yourself, make your career goals clear, and build a promotion readiness plan with your leader. Doing so holds you *both* accountable.

The Best Feedback Stings a Little

Sometimes, I think back to moments when I received tough feedback, and I still cringe a little. (Even a decade later!) It's because the feedback was spot-on.

As a product leader, you're only as strong as your ability to influence and rally those around you. That makes specific feedback exceptionally important, so you can understand the gap between how you think you show up and how others think you show up. This includes feedback from your direct team, your peers, your manager, and yes, even difficult partners. Establish feedback loops about your performance that give you specific skills to work on, even if it stings a little bit. The sting just signals a bit of discomfort, but it's also the catalyst for growth.

Tours of Duty

Product life cycle stages move slower than most PMs expect. You could easily spend 3–5 years working on a product without advancing it to the next stage. There's nothing inherently wrong with that, but competitive product leaders have fluency across the entire product life cycle. That includes 0–1 product management all the way through sunsetting solutions. If you're not gaining exposure to all parts of the product life cycle, look for ways to fill these gaps. Sometimes, you need tours of duty on other teams or companies to build end-to-end exposure. Of course, there's nothing wrong with being a specialist either, but if you're craving breadth, try the following:

- Ask your leader for stretch projects or short-term assignments.
- Rotate to support another product within the same company.
- Volunteer for special projects at your current company or within your community.
- Find a role externally that provides a chance to learn new skills.

LEARN MORE

ADDITIONAL RESOURCES FOR PEOPLE LEADERSHIP

The world of coaching and leadership is vast, and a short list of resources won't do it justice. But if you want to start somewhere, start with these first:

- *The Making of a Manager: What to Do When Everyone Looks to You* by Julie Zhou
- *Empowered: Ordinary People, Extraordinary Products* by Marty Cagan and Chris Jones
- *We Need to Talk: A Survival Guide for Tough Conversations* by Joshua Graves

Thought leaders to follow include Melissa Perri, Marty Cagan, Christian Idiodi, and Shreyas Doshi

Pulling It All Together

The digital product field is chock-full of rapidly shifting technology, compelling frameworks, and viral corporate content, but often it's the good old-fashioned virtues that matter most. These virtues are having courage to make tough calls, pursuing excellence, modeling integrity, following the golden rule, and being willing to grow. These are the ingredients that fuel longevity and career resilience. At the end of the day, your success won't come from pursuing a title or bumps on the org chart. You'll feel it when you've developed the confidence to inspire, motivate, and lead through good times and stormy seas. Keep growing, keep leading, and the rest will follow.

CHAPTER 12

The Future of Product Management and You

Corporate longevity ain't what it used to be. In 1958, the average lifespan of an S&P 500 company was 61 years. Fast forward to the 2020s, and today the average lifespan is less than 18 years. And startups fare even worse; most reports indicate that 90% of startups end up failing.[1,2]

Woof. What's causing so many businesses to collapse?

Plenty of variables—global crises, volatile interest rates, regulatory changes, evolving consumer preferences, and relentless technological disruption. Of course, there will always be winners and losers in business, and companies succeed or fail based on their ability to innovate, keep customers satisfied, and outpace the competition. But it's the pace of change, not change itself, that's forcing transformation at unprecedented speed and keeping C-suite leaders up at night.

As a product creator, your career isn't immune to these forces. Years ago, the buzz was cloud services and design thinking, now it's AI products and growth loops, and in the future it will be _______________.*[3] The skills and knowledge that got you hired last time may not keep you employed five years from now. Just like a business, you, too, must continuously evolve your value proposition to earn a seat at the table. You can't preach product innovation but let your own skills grow stale. Instead, future-proof your career by asking the questions that matter most:

- How can I stay relevant?
- How do I stay resilient?
- What's next for me?

1 McKinsey & Company, "McKinsey on Finance," no. 84 (December 2023), www.mckinsey.com/capabilities/strategy-and-corporate-finance/our-insights/mckinsey-on-finance/mckinsey-on-finance-number-84

2 Brian Heater, Mary Ann Azevedo, and Zack Whittaker, "Remembering the Startups We Lost in 2023," TechCrunch, December 30, 2023, accessed November 12, 2025, https://techcrunch.com/2023/12/30/remembering-the-startups-we-lost-in-2023/

3 *Insert predictions for the next seismic trends here!

Strategies for Career Longevity

Long careers in product management aren't built on annual conferences. Maintaining relevance is more like tending to a garden; it's a continuous process that requires a growth mindset, rolling up your sleeves, and hands-on experience. Here are a few strategies to set yourself up for success.

Strategy 1: Focus on Adding Value, Not Title

It's no secret that companies often rework job families, functions, and org charts to become more efficient and profitable over time. Once popular roles, like Scrum Masters and dependency managers, are no longer as prevalent. Meanwhile, newer roles in product-led growth, product marketing, AI product management, and product operations have emerged. Evolving roles create pressure on product teams to reimagine responsibilities and adapt to new teaming styles. And yes, sometimes it feels like *more work*, but here's the best advice when this happens: work *with* change rather than *resisting* it.

I'm not saying role changes are easy; I'm saying they are a reality. (Heck, what we know as product management today will surely look different five to ten years from now.) But when your company changes career ladders, functions, or team structure, don't be a bottleneck; focus on learning skills that *add value* to your organization and adapt as quickly as possible, instead. This brings us to the next strategy.

Strategy 2: Be an Early Adopter

You can't ignore the tectonic shift happening under your feet right now. AI is dramatically disrupting industries, companies, and people's paychecks. It also levels the playing field between experienced and inexperienced product creators. Companies of the future will expect more from product managers than synthesizing insights into vision and orchestrating team activities across the product life cycle. They'll also expect PMs to be more *hands-on*, co-create prototypes, test solutions, train models, and use data directly. Here are three ways you can start embracing this change now:

- **Build data fluency.** If you've conquered data literacy (the ability to interpret and leverage data in context), then focus on building data fluency. Data fluency is the ability to manipulate and work with data sets and analytical tools to derive insights and make strategic decisions.

- **Embrace AI.** Experiment with AI tools to accelerate product life cycle activities. If you're just getting started, use AI tools to analyze customer support patterns, design early prototypes, or even fine-tune your requirements doc.
- **Start small but start *somewhere*.** Whether it's AI or machine learning, you need a sandbox to start learning and iterating. Sign up for vibe-coding workshops or partner with engineering on a small proof-of-concept to get hands-on exposure.

Strategy 3: Seek and Apply an Outside Perspective

As a hiring manager, I often ask candidates to tell me what they read, who they follow, or how they keep a pulse on industry and technology insights. It's astounding how many of them don't reply with a specific resource. In an instant, I can tell they aren't serious about product craft. On the other hand, those who reply with specific influencers, books, podcasts, etc. perform stronger during case studies and score higher on leadership assessments.

Here's why. When you only learn about product management, ways of working, and technology from people around you, you're stuck in an echo chamber. That control tower between your ears needs inspiration from thought leaders, entrepreneurs, and industry peers to avoid repeating the same mistakes and limiting your imagination. Even small doses of wisdom make a big difference. Here's how to make it systematic:

- **Spend 30 minutes a week learning from outsiders.** Whether it's a podcast, an article, a book, or a mentorship call, keep learning. Figure 12.1 offers the starter kit for top influencers and communities to follow.
- **Apply one new idea every quarter.** Did you learn how a B2B company unlocked a new growth loop? Did someone share their framework for stronger evals? What about that leadership insight you read about? Take one thing you learned and adapt it to a real-life use case every quarter. And score some bonus points with your team (and boss) by sharing what you learned after trying something new.

The PM's Starter Kit for Staying Inspired

- **Lenny Rachitsky:** host of Lenny's Podcast & newsletter author with over one million subscribers
- **Marty Cagan and Christian Idiodi:** authors, speakers, and product experts at Silicon Valley Product Group
- **Dan Olsen:** author and leader of the Lean Product Meetup community
- **Dr. Marily Nika:** AI product expert, author, and educator
- **Teresa Torres:** product discovery expert, author, and podcast host
- **Melissa Perri:** product coach, author, and podcast host
- **Paweł Huryn:** author of The Product Compass newsletter
- **Mind the Product:** community with 300K followers worldwide
- **Women in Product:** community with 75K followers worldwide

FIGURE 12.1

Follow the experts and sharpen those skills. You'll be glad you did!

Interviewing Like a Pro

We can't discuss future-proofing strategies without touching on interview skills. Most product manager hiring journeys involve a multistep gauntlet designed to pressure test multiple skill sets, assess leadership behaviors, and ultimately sniff out mediocre PMs from exceptional ones. The bootcamps, credentials, and prior experience might demonstrate passion on your résumé, but interview performance is where you sink or swim.

During the hiring process, many companies switch up interview formats to assess behavioral, analytical, product design, technical, and team fit criteria. While the formats differ across industries, most hiring managers generally need evidence that confirms the following information before extending you an offer:

- You *already* have experience creating, shipping, and optimizing products.
- You can strategize *and* execute.

- You're fluent across business, customer, technology, design, and data domains.
- You can operate independently on Day 1.
- You can influence anyone with clear communication and leadership behaviors.
- People enjoy working with you.

Even though PM interviews are designed to mimic real-life scenarios, they're often harder than the real job because the questions are intentionally ambiguous, the clock is ticking, and the stakes are higher. Many candidates look exceptional on paper, but when it's showtime, they struggle to show breadth of skills, answer with the right level of depth, and maintain enthusiasm under pressure.

As a hiring manager, I've conducted countless interviews, and noticed one behavior that sets great PM candidates apart from mediocre ones: stronger PMs deeply invest in interview prep. If you're willing to put in the work (and I'm talking 5+ hours of prep), you can conquer the pressure-cooker and influence with confidence. The chart below illustrates key behaviors to help you shine brighter during interviews.

LEARN MORE

ESSENTIAL RESOURCES FOR NAVIGATING THE PM INTERVIEW PROCESS

Fantastic resources exist to help you navigate the world of product management hiring practices. Start with these first.

- ***Cracking the PM Interview: How to Land a Product Manager Job in Technology*** **by Gayle Laakmann McDowell and Jackie Bavaro**
- **IGotAnOffer for interview guides, mock interview prep, and résumé feedback**
- **Levels.fyi for salary details you can use when negotiating**
- **r/ProductManagement on Reddit for support and advice (It's lively, check it out!)**

THE SMART PM'S CHECKLIST FOR EFFECTIVE INTERVIEWS

Before the interview:

- ☐ Deeply research the company, the business model, and competitors.
- ☐ Hypothesize different personas or customer segments for the company.
- ☐ Explore and critique the public-facing digital experience.
- ☐ Practice sketching or vibe-coding improvements (just in case).
- ☐ Write down real-life stories that reflect your PM skills in action.
- ☐ Prepare and rehearse Q&A aloud until you're clear *and* concise.
- ☐ Ask the recruiter for advice on what to expect.

During behavioral or team fit interviews:

- ☐ Reference insights from company research (i.e., share product critique, refer to core business model, share insights about the executive team, etc.).
- ☐ Leverage concise, effective storytelling that ends in a concrete result.
- ☐ Show awareness of product management trends (i.e., different philosophies, roles, or leaders).
- ☐ Articulate your brand and superpower (i.e., share your top skills, peer feedback, etc.).
- ☐ Offer leadership examples that show how you went beyond basic expectations.
- ☐ Be clear on what you want next (i.e., special interests, grow into a leader role, etc.).

continues

THE SMART PM'S CHECKLIST FOR EFFECTIVE INTERVIEWS (continued)

During product design case interviews:

- ☐ Always start with why, clarify the opportunity, and define business success up front.
- ☐ Think aloud and cover the *entire* product development process, including core insight, customer segments, needs, value prop, requirements, design, technology, and data architecture.
- ☐ At each step, flare on ideas and then narrow on a choice. Declare assumptions when moving forward and making trade-offs.
- ☐ Propose a cohesive vision, strategy, and roadmap and show how you'd deliver it.
- ☐ Reference specific frameworks, skills, and techniques to handle product discovery, product delivery, go-to-market, and scale.
- ☐ Reference specific techniques to influence and lead stakeholders.

After the interview:

- ☐ Send a personalized thank-you note.
- ☐ Send examples of your work (i.e., share a portfolio, product critique, writing sample, etc.).

Building Inner Resilience

So far, I've shared ways to future-proof your career by focusing on external factors, including technology adoption, skills development, and interview behaviors. But career longevity doesn't just require relevance, you need *resilience*, too. Resilience is what enables you to stay motivated, weather setbacks, and cultivate perspective as your workplace identity evolves. Of course, building resilience looks different for everyone, but here are a few universal practices to help you navigate seasons of change:

- **Build a team of trusted advisors.** Breakthroughs can happen when someone spots potential in you that you don't see in yourself. (That's exactly how I landed in product management.) Whether it's a mentor, family member, professional coach, or peer, build a small network of people who truly get you. Call on them for advice and support.
- **Celebrate your growth.** It might sound cheesy, but take time every month, quarter, and year to zoom out and celebrate your development. While you're busy fighting ambiguity and blockers every day, you're also sharpening skills that prepare you for the next challenge. Reflecting on your growth and lessons can boost your confidence and self-belief.
- **Prioritize life outside of work.** Sleep, family, nature, friendship, fun hobbies. These ingredients fuel the creativity you need to solve hard problems, lead well, and maintain consistency and perspective when work goes sideways. Take time to protect the most important resource you have—yourself.

Deciding What's Next

Product managers often fixate on sprints and velocity, but your career is something that unfolds over decades. Of course, you can craft a ten-year plan if that feels right for you, but progress only happens one step at a time. If you're feeling stuck on what to do next, ask yourself three questions to sort it out:

- What matters most to me today? (Stability, growth, mastery, money, culture, something else?)
- What are my current strengths and gaps? (Your Chapter 2, "The Product Management Skills Assessment," has answers.)
- What's one concrete move I can make *right now*?

These questions are simple, but effective. The truth is that you waste so much time overthinking and analyzing, but the best learning comes from doing. Make a plan, try it out, and adjust. Then rinse and repeat.

Pulling It All Together

This chapter began with sobering statistics about corporate mortality in the modern era, but it's time to consider the optimist's view. The relentless pace of change is fueling innovation at scales previously impossible, and product managers are in the driver's seat. The last decade produced extraordinary product breakthroughs that reshaped how the world learns, heals, and connects. We now have free digital learning platforms that democratize education and spread knowledge to millions of learners that otherwise wouldn't have access. Telemedicine now enables remote care for those who can't reach doctors. Evolutions in video, voice, and texting tools make it easier to express yourself, collaborate, and connect with loved ones.

As a product builder and innovator, you get to shape the next breakthrough. Whether you're supporting a small B2B SaaS startup, launching the next big healthcare innovation, or anything else, the world needs people who can translate complex needs into working solutions that do good.

This book is your companion as you build a brighter future. I hope it inspires you, energizes you, and gives you the confidence to know how to think and what to do. And above all, I hope you use it to build products that matter.

APPENDIX: GUIDE TO THE METHODS IN THIS BOOK

The table below lists all the playbook methods included in this book. Refer to the "focus" column to determine when and why you should use a particular method.

#	Playbook Method	Focus	Average Time
1	Leading a Business Strategy Workshop	What is the company's goal and corporate strategy right now? What market does the company want to capture? Why?	3–4 hours
2	Building Problem/Opportunity Statements to Find a Hunch	What's the strategic problem or opportunity to investigate and why is it important?	3–5 days
3	Mapping Competitors	Who is the competition and why do they matter?	4–8 hours
4	Building a Research Plan	What is my plan to learn about our customer?	2–4 hours, plus time for feedback
5	Eliciting Needs via Customer Interviews	What are my customers' functional needs, emotional needs, and biggest pain points? How do they solve these problems today?	1 week, plus time for synthesis
6	Sharing Research Readouts	What did we learn about customers and what should we do about it?	1–2 days
7	Generating Solution Ideas	What solutions solve the target customer's needs?	4–5 hours
8	Testing Product Concepts	What is our test plan to validate/invalidate early ideas?	5–10 days
9	Building a Problem-Solution Fit Map	What's a quick way to pressure test initial solution scope?	2–8 hours

continues

#	Playbook Method	Focus	Average Time
10	Assessing Technical Feasibility	What will it take to make my product actually work? How much effort will it take to build it? How can I future-proof my product?	1–2 weeks
11	Projecting Impact & Defining Success Metrics	How would this solution benefit our business? How will we measure success?	5–8 hours
12	Defining a Product Vision	What is our product vision?	1 week, plus time for feedback
13	Defining a Product Strategy	What's our strategy to succeed in-market?	1–3 days, plus time for feedback
14	Building a Roadmap	What will we launch, when will we launch it, and what impact will it have?	2–3 days, plus time for feedback
15	Building a Backlog	How do I break down the roadmap into work?	2 hours per feature
16	Crafting Product Requirements	What does the team need to be successful?	2–3 hours per feature, plus time to refine
17	Selecting a Pricing Model	What are customers willing to pay?	2 weeks
18	Building a Go-to-Market Plan	What must be true to "Go to Market"?	2 weeks, plus time to execute
19	De-risking the Release	How can we ensure a smooth release?	2 hours, plus time to implement
20	Optimizing with A/B Tests	How can we assess the impact of one experience over another?	1 day to plan, plus time to implement
21	Rehabbing a fledgling product	What should I do when my solution isn't driving value or accomplishing goals?	1 week, plus time to implement changes

INDEX

Numbers

A

D

E

F

G

H

I

J

K

L

M

N

O

P

Q

R

S

T

U

V

W

Z

ACKNOWLEDGMENTS

Writing a book on digital product management means chasing a field in constant motion. This book is my attempt to distill the enduring principles of product craft while keeping pace with the tools, roles, and technologies shaping how product work happens today. It's a collection of concepts, principles, and methods for everyday product builders making sense of the messy middle, and I couldn't have written it without the collective wisdom from thought leaders, mentors, and practitioners across business, product, design, technology, and leadership. What follows is my attempt to thank those who shaped it most.

Iron sharpens iron, and the best product teams are built by people willing to sharpen others. The experts featured in this book are exactly those people: practitioners who work alongside product teams and elevate everyone around them. My deepest thanks go to Des Traynor, Teresa Torres, Ally Hwang, Vito Covalucci, Derek Schwartz, Leon Feuerstein, Chetan Kandhari, and Dan Olsen for sharing your wisdom with readers.

I've had the immense privilege of growing alongside brilliant peers and mentors throughout my career, including Betsy Bland, Joanna Snead, Kristina Bailey, Derek Schwartz, Ally Hwang, Clay Johnson, and Mindy Seibold. Each of them shaped how I think, how I lead, and many of the chapters in this book. I want to especially recognize Katie Holder, my trusted advisor, friend, and champion of this book long before I wrote a single word.

Reviewing a manuscript is a significant responsibility. Clay Johnson, Karim Rios, Byron Sullivan, Derek Schwartz, John Hoyt, and Christian Crumlish each gave their time and honesty to early drafts of this book. It's a much sharper book because of their feedback. In that same spirit of professional generosity, writing a foreword is a massive undertaking. I feel incredibly fortunate that John Cutler lent his unique perspective and guidance to set the stage for readers.

After a decade-plus of reading Rosenfeld Media's books, publishing with them is a full-circle moment. Working with the Rosenfeld team has been a genuine privilege. My editor, Marta Justak, coached me through every stage of the writing process, never let me lose confidence in what I was building, and made each chapter stronger. And Lou Rosenfeld has created an environment where authors thrive, and ideas have room to grow. Thank you, Lou, for reading my proposal and believing this leap was worth taking.

I'm blessed to have an incredibly supportive family. Mom, Dad, Bill, Gladys, and Tony—thank you for cheering me on, making me laugh, and supporting me, even though my job remains a mystery to you. You keep me grounded in what matters most.

And of course, a million thank-yous to my love, Eric, for riding the emotional rollercoaster of book writing with me. You gave me time and space to write, talk it out in the kitchen, revise, read it aloud, revise again, and finally send in each chapter. Thank you for keeping me sane and well-fed throughout this process. I love you, forever.

Rosenfeld

Dear Reader,

Thanks very much for purchasing this book. There's a story behind it and every product we create at Rosenfeld Media.

Since the early 1990s, I've been a User Experience consultant, conference presenter, workshop instructor, and author. (I'm probably best-known for having cowritten *Information Architecture for the Web and Beyond*.) In each of these roles, I've been frustrated by the missed opportunities to apply UX principles and practices.

I started Rosenfeld Media in 2005 with the goal of publishing books whose design and development showed that a publisher could practice what it preached. Since then, we've expanded into producing industry-leading conferences and workshops. In all cases, UX has helped us create better, more successful products—just as you would expect. From employing user research to drive the design of our books and conference programs, to working closely with our conference speakers on their talks, to caring deeply about customer service, we practice what we preach every day.

Please visit rosenfeldmedia.com to learn more about our **conferences**, **workshops**, **free communities**, and **other great resources** that we've made for you. And send your ideas, suggestions, and concerns my way: louis@rosenfeldmedia.com

I'd love to hear from you, and I hope you enjoy the book!

Lou Rosenfeld,
Publisher

RECENT TITLES FROM ROSENFELD MEDIA

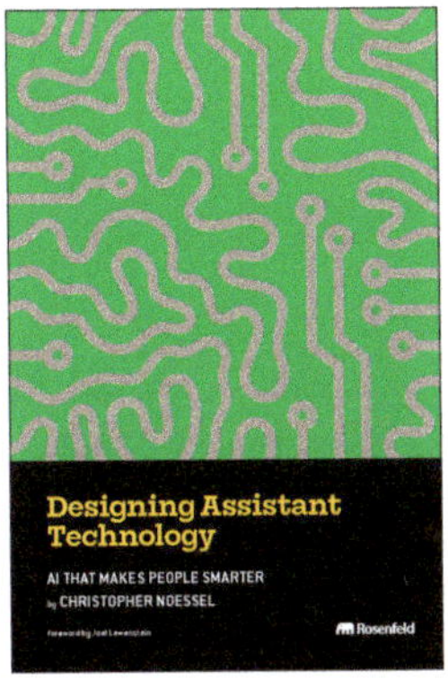

Get a great discount on a Rosenfeld Media book:
visit **rfld.me/deal** to learn more.

SELECTED TITLES FROM ROSENFELD MEDIA

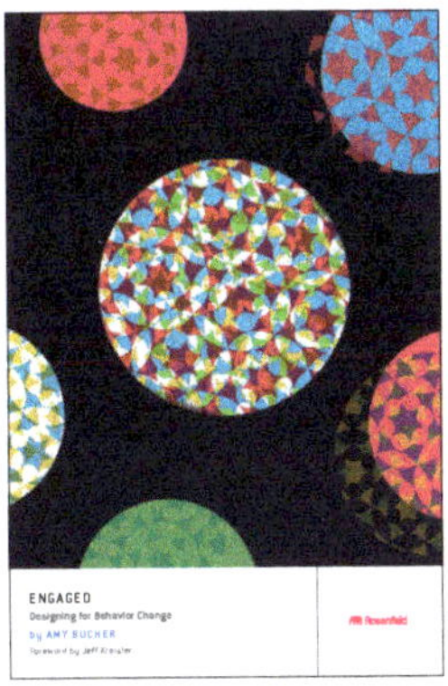

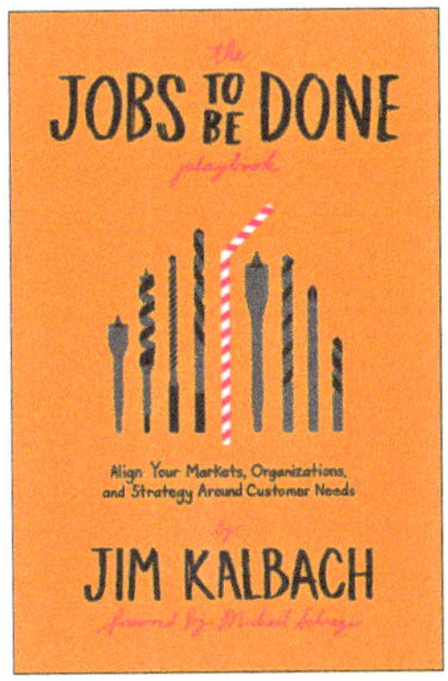

View our full catalog at **rosenfeldmedia.com/books**

ABOUT THE AUTHOR

Julia Barham is a product executive, patented inventor, and veteran of the full product journey—from zero-to-one launches to scaling market-leading solutions across Fortune 100 companies. She has built inaugural product teams, led B2C and B2B products, and run digital experimentation engines that generated multimillion-dollar revenue gains.

Julia's work has earned top-tier recognition for innovation and design, but her true passion is helping teams cut through corporate complexity and bridge the gap between ambitious strategy and daily execution—the space where most business leaders and product teams struggle.

She's a firm believer that great product leaders are built, not born, and she has mentored countless rising product leaders across end-to-end product management topics. Julia wrote this book to help everyday practitioners move faster, lead with clarity, and build products that actually matter.

Julia lives in Cleveland, Ohio, with her husband Eric and beloved English Springer Spaniel, Sally.